Manual of Water Supply Practices

M69

Inland Desalination and Concentrate Management

I088188

American Water Works Association

Manual of Water Supply Practices—M69

Inland Desalination and Concentrate Management

Editorial Manager – Book Products: Melissa Valentine
Technical Editors: Dianne Rose and Suzanne Snyder
Cover Art/Technical Illustrator: Michael Labruyere
Production: Innodata
Manuals Specialists: Sue Weikel and Willadee Hitchcock

Library of Congress Cataloging-in-Publication Data
Names: He, Charlie (Qun), author. | Bond, Rick G., author. | American Water Works Association, issuing body.
Title: Inland desalination & concentrate management / by Charlie (Qun) He and Rick G. Bond.
Other titles: Inland desalination and concentrate management | AWWA manual ; M69.
Description: First edition. | Denver, CO : American Water Works Association, [2019] | Series: Manual of water supply practices ; M69 | Includes bibliographical references and index.
Identifiers: LCCN 2018061027 | ISBN 9781625763303
Subjects: LCSH: Saline water conversion.
Classification: LCC TD479 .H4 2019 | DDC 628.1/67--dc23
LC record available at https://lccn.loc.gov/2018061027

Printed in the United States of America

ISBN 978-1-62576-330-3 eISBN-13 978-1-61300-502-6

American Water Works Association

American Water Works Association
6666 West Quincy Avenue
Denver, CO 80235-3098
awwa.org

Contents

List of Figures, v

List of Tables, ix

Preface, xiii

Acknowledgments, xv

Chapter 1 Overview of Inland Desalination 1
Brackish Water Sources, 2
Status of Desalination, 2
Concentrate Management Options, 7
Manual Overview, 11
References, 12

Chapter 2 Water Quality and Planning Strategies.............................13
Source Water Quality Affecting Desalination, 14
Product Water Quality Goals, 18
Water Quality Analysis And Monitoring, 21
RO and EDR Performance Modeling, 23
Treatment Levels, 23
Types of Brackish Water Desalination Technologies, 25
Types of Concentrate Management Technologies, 26
Planning Strategies, 28
Conclusions, 32
References, 32

Chapter 3 Brackish Water Desalination......................................35
Pressure-Driven Membrane Desalination, 36
Reverse Osmosis Pretreatment, 47
Reverse Osmosis Post-treatment, 50
Electrodialysis Desalination, 51
Electrodialysis/Electrodialysis Reversal Pretreatment, 65
Electrodialysis/Electrodialysis Reversal Post-treatment, 67
References, 68

Chapter 4 Discharge Options for Concentrate Disposal 71
Discharge to Surface Water, 72
Discharge to Sewer, 75
Deep-well Injection, 76
Evaporation Ponds, 80
Enhanced Evaporation Systems, 84
Irrigation, 87
References, 91

Chapter 5 **Enhanced Recovery and Zero Liquid Discharge. 93**
Enhanced Recovery Processes, 93
Zero Liquid Discharge, 105
References, 117

Chapter 6 **Cost of Desalination and Concentrate Management 119**
Overall Cost Considerations, 120
Methods for Estimating Class 4 and Class 5 Construction Costs, 121
Construction Costs for Major Processes Used in Desalination and Concentrate Management, 122
Annual Costs, 127
References, 130

Chapter 7 **Regulatory, Safety, Operational, and Environmental Issues 131**
Current Concentrate Management Issues, 131
Regulatory Issues, 132
Federal Regulations, 133
State Regulations, 138
Relevant International Regulations, 140
Future Permitting Considerations, 141
Environmental Impacts, Operations, and Permitting Requirements, 141
References, 152

Chapter 8 **Case Studies 155**
City of Chandler Ocotillo Brine Reduction Facility, 155
City of Palm Coast Zero Liquid Discharge Reverse Osmosis Concentrate Management, 160
Irwin Water Works, Army National Training Center, Fort Irwin, Calif., 164
Fargo Wastewater Treatment Facility, Effluent Reuse Facility, 167
Kay Bailey Hutchison Desalination Plant, El Paso, Tex., 171
Zero Liquid Discharge Water Treatment Facility, 175
Laguna County Sanitation District Deep-Well Concentrate Disposal, 179
Chino Desalter Authority Concentrate Reduction Facility, 182

Chapter 9 **Technological Trends, Salt Recovery, and Beneficial Uses. 187**
Industry Trends, 188
Innovations in Membrane Technology, 188
Innovations in Design and Operation, 190
Innovative Processes, 192
Energy Efficiency Enhancement, 194
Improved Salt Recovery and Beneficial Uses of Concentrate, 196
References, 202

Index, 205

List of AWWA Manuals, 213

Figures

1-1 Depth to brackish groundwater in the United States, 3
1-2 Total worldwide installed desalination capacity by feedwater category, 3
1-3 Total worldwide installed desalination capacity by user type, 4
1-4 Cumulative number of US municipal desalination plants, 5
1-5 Cumulative installed capacity of US municipal desalination plants, 5
1-6 Number of US municipal desalination plants by state as of 2017, 6
1-7 Number of US municipal desalination plants installed by state in different time periods, 7
1-8 Concentrate disposal options for US municipal desalination plants, 10

2-1 Treatment and disposal levels for inland desalination and concentrate management, 24
2-2 Three strategic focuses for inland desalination and concentrate management, 27

3-1 Composite polyamide membrane structure, 36
3-2 Basic concept of natural osmosis and reverse osmosis, 37
3-3 Schematic representation of a spiral-wound membrane module, 40
3-4 Eight-inch RO membrane modules being loaded into a membrane pressure vessel, 41
3-5 Schematic representation of a two-stage RO process, 41
3-6 Schematic representation of a two-stage RO process with a turbocharger energy recovery device, 43
3-7 Energy recovery device installed in a brackish water desalination facility, 44
3-8 Horizontal cartridge filter vessels used as RO pretreatment, 49
3-9 The Taunton River desalination plant in Brockton, Mass., uses integrated membrane systems to treat a brackish surface water supply., 50
3-10 Water treatment spectrum, 52
3-11 Electrodialysis process schematic, 52
3-12 Structure of a cation membrane, 53
3-13 Structure of an anion membrane, 54
3-14 Spacers, 54
3-15 Baffled flow pattern created by spacer, 55
3-16 An ED system with three hydraulic and three electrical stages, 56
3-17 Hydraulic staging within a single membrane stack, 56
3-18 Membrane stack with two electrical stages and three hydraulic stages, 57
3-19 Raw water concentration and current density, 58
3-20 EDR stack shorting, 59

4-1 Class I injection well schematic (number of casing stages varies), 78
4-2 Salt gradient solar pond schematic, 86
4-3 Plant salt tolerance classifications, 90

5-1 Schematic of pellet-softening process, 95
5-2 Schematic of an EDR system, 96
5-3 Process flow diagram of an EDR system, 97
5-4 Schematic of lime/soda softening plus secondary RO or EDR, 98
5-5 EDM process schematic and picture of full-scale stack, 99
5-6 SPARRO process schematic, 99
5-7 Integrated EDR-SPARRO process schematic, 100
5-8 Vacuum-enhanced direct contact membrane distillation process schematic, 101
5-9 Membrane brine concentrator process schematic, 102
5-10 Concept of the VSEP, 103
5-11 Example of the VSEP configuration, 103
5-12 Schematic of the HERO™ process, 104
5-13 Overview of the CCRO process, 104
5-14 Brine concentrator process schematic (typical), 107
5-15 Crystallizer (left) and brine concentrators (typical), 110
5-16 Crystallizer process schematic (typical), 110

7-1 USEPA classification of waste streams, 134
7-2A Solid residuals TENORM decision tree, 137
7-2B Liquid residuals TENORM decision tree, 138
7-3 Surface water discharge, 146

8-1 CHRO process flow schematic (prior to spring 2014), 156
8-2 OBRF process flow schematic (after spring 2014), 159
8-3 ZLD process schematic, 161
8-4 Screenshot of ZLD SCADA system, 164
8-5 Facility location map, 165
8-6 Irwin Water Works process flow diagram, 166
8-7 Facility location map, 168
8-8 Fargo wastewater reuse project schematic, 168
8-9 Kay Bailey Hutchison Desalination Plant location (El Paso County, Tex.), 171
8-10 Kay Bailey Hutchison plant desalination process schematic, 172
8-11 Kay Bailey Hutchison plant RO trains, 172
8-12 ZLD treatment schematic, 175
8-13 Falling film vapor compression evaporator, 176
8-14 One of the triple-lined evaporation ponds used for final disposal of concentrate from the thermal evaporator, 178
8-15 Facility location map, 179
8-16 LCSD treatment schematic, 180

8-17 Facility location map, 183
8-18 Chino II CRF process schematic, 183
8-19 Pellet reactors (left) and secondary RO (right) at the Chino II CRF, 185

9-1 General schematic of the thermo-ionic process, 193
9-2 Selective salt recovery from sodium carbonate crystals, 199

This page intentionally blank.

Tables

1-1 Number of different US municipal desalination membrane processes as of 2017, 6
1-2 Concentrate management options, 8
1-3 Number of states using the disposal options for municipal desalination concentrate, 9
1-4 States using various disposal options and number of plants in each state, 9

2-1 Example of brackish water quality variations, 15
2-2 Examples of drinking water quality goals, 19
2-3 Examples of short-term salinity goals, 20
2-4 Regulations in California for groundwater recharge into potable aquifers, 21
2-5 Example of product water quality criteria for direct potable reuse applications, 22
2-6 Reverse osmosis and nanofiltration design computer models, 24
2-7 Types of desalination technologies, 27

3-1 Impact of operational changes on removal efficiency and pressure, 39
3-2 Methods used for balancing flux between stages, 42
3-3 Typical water impurities and corresponding pretreatments, 65

4-1 Advantages and disadvantages of discharge of concentrate to surface water, 74
4-2 Advantages and disadvantages of discharge of concentrate to sewer, 76
4-3 Advantages and disadvantages of deep-well injection, 79
4-4 Advantages and disadvantages of discharge of concentrate to evaporation ponds, 83
4-5 Advantages and disadvantages of spray systems, 84
4-6 Advantages and disadvantages of WAIV, 85
4-7 Advantages and disadvantages of salt-gradient solar ponds, 87
4-8 Examples of crop salt tolerance, 90
4-9 Advantages and disadvantages of irrigation with concentrate, 90

5-1 Influence of feed TDS and maximum concentrated brine effluent TDS on brine concentrator recovery, 108
5-2 Influence of flow on crystallizer operating costs (example), 111
5-3 ZLD treatment cost example, 112
5-4 RO treatment cost example at 80% recovery, 113
5-5 RO+ZLD combined treatment cost example at 80% RO recovery, 113
5-6 Sample RO and ZLD treatment cost comparison at 80% RO recovery, 114
5-7 RO treatment cost example at 90% recovery, 114
5-8 RO+ZLD combined treatment cost example at 90% RO recovery, 114

5-9 Sample RO and ZLD treatment cost comparison at 90% RO recovery, 114
5-10 Comparison of RO+ZLD combined treatment costs at 80% versus 90% RO recovery, 115

6-1 Summary of major treatment cost by category, 120
6-2 Guidelines for cost estimate classes, 121
6-3 Examples of cost add-ons, 123
6-4 Cost estimation example, 124
6-5 Typical MF/UF construction cost, 124
6-6 Typical brackish water NF/RO membrane construction cost, 125
6-7 Typical seawater NF/RO membrane construction cost, 125
6-8 Equipment costs for reactor clarifier for installation in 15-ft deep concrete or steel wall, 126
6-9 Pellet softener equipment costs including sand wash/feed pump and control, 126
6-10 Deep-well injection costs from recent installations, 127
6-11 MVC brine concentrator equipment costs, 128
6-12 MVC crystallizer equipment costs, 128
6-13 Chemical cost example based on 10 mgd brackish water RO plant, 129
6-14 Manpower cost example based on 10 mgd brackish water RO plant, 129
6-15 Power consumption example based on 10 mgd brackish water RO plant, 130

7-1 Disposal methods in the United Kingdom, 132
7-2 Comparison of raw water and concentrate, 143
7-3 Comparison between salinities of seawater and membrane-filtered concentrates, 143
7-4 State of Florida potable water reverse osmosis and nanofiltration systems that discharge to surface waters or ocean outfalls, 145

8-1 Case study facilities, 156
8-2 ZLD water quality parameters, 162
8-3 UF membrane design criteria, 162
8-4 Concentrate ZLD design flows, 163
8-5 EDR system design criteria, 166
8-6 EDR water qualities, 167
8-7 UF membrane characteristics, 169
8-8 RO unit parameters, 169
8-9 ERF water quality, 170
8-10 RO unit parameters, 173
8-11 KBH water quality (February 2015), 173
8-12 RO system characteristics, 176
8-13 Brine concentrator system, 177
8-14 Evaporation ponds, 177
8-15 MBR system characteristics, 181
8-16 RO system characteristics, 181

8-17 WDR permit limits for discharge, 181
8-18 Monitoring requirements for deep well, 182
8-19 Primary RO characteristics, 184
8-20 Water quality summary, 184
8-21 Secondary RO characteristics, 184
8-22 Anticipated chemical usage, 185

9-1 Water quality before and after pelletized treatment, 197
9-2 List of recoverable salts and their potential application, 201

This page intentionally blank.

Preface

With population and economic growth stressing global freshwater supplies, desalination of brackish water and the appropriate management and disposal of concentrate are increasingly attractive options. In the past few decades, the number and capacity of US municipal desalination plants increased rapidly. More than one-third of these facilities are in inland areas. Many concentrate management and disposal options were proven at these facilities, including surface water discharge, discharge to sewer, deep-well injection, land application, discharge to evaporation ponds, near-zero liquid discharge, and zero liquid discharge.

Because of rapid growth, the need for a manual is urgent. The industry has accumulated enough knowledge and experience to provide guidance. After several years of diligent coordination, the AWWA Water Desalting Committee proudly presents AWWA Manual M69, *Inland Desalination and Concentrate Management*.

This manual provides technical and planning guidance for inland water utilities (public and private) that are operating, developing, or considering brackish water desalination. It presents practical information on inland desalination, concentrate treatment technologies and management strategies, permitting procedures, environmental impacts, costs, safety, and more.

The concentrate management topic of this manual is also relevant to direct and indirect potable reuse practices where nanofiltration and reverse osmosis might be utilized. It can also apply to produced water treatment, industrial ultrapure water system brine management, and cooling water blowdown management.

Charlie (Qun) He
Chair
M69 *Inland Desalination and Concentrate Management*

This page intentionally blank.

Acknowledgments

Chapter authors were recruited for their technical expertise from various fields, including universities, water utilities, public outreach companies, and engineering companies.

Manual Participants

Steering Committee

Q. He, Carollo Engineers, Phoenix, Ariz.

R.G. Bond, Black & Veatch, Kansas City, Mo.

D.R. Stewart, Stewart Environmental Consultants, Fort Collins, Colo.

S. Sethi, Carollo Engineers, Sarasota, Fla.

S.D.N. Freeman, Black & Veatch, Kansas City, Mo.

V.S. Frenkel, Greeley and Hansen, San Francisco, Calif.

S.A. Weikel, AWWA, Denver, Colo.

A. Gerling, AWWA, Denver, Colo.

Authors

Chapter 1: Overview of Inland Desalination
Greg Wetterau and Mike Mickley

Chapter 2: Water Quality and Planning Strategies
Charlie (Qun) He, Arun Subramani, Greg Wetterau, and Vasu Veerapaneni

Chapter 3: Brackish Water Desalination
Val S. Frenkel and Greg Wetterau

Chapter 4: Discharge Options for Concentrate Disposal
Rick Bond and Sandeep Sethi

Chapter 5: Enhanced Recovery and Zero Liquid Discharge
Brent Alspach and Graham Juby

Chapter 6: Cost of Desalination and Concentrate Management
Jorge T. Aguinaldo and Rick Bond

Chapter 7: Regulatory, Safety, Operational, and Environmental Issues
Dave Stewart, Fred Bloetscher, and Melanie Goetz

Chapter 8: Case Studies
Howard E. Steiman, Qigang Chang, Charlie (Qun) He, Phil Locke, Ryan Popko, Jim Hogan, Fred Greiner, Jason Yoshimura, Bill O'Neil, Greg D. Wetterau, Shane Walker, Malynda Cappelle, Tom Seacord, Graham Juby, and Brandon Yallaly

Chapter 9: Technological Trends, Salt Recovery, and Beneficial Uses
Ali Sharbat, Charlie (Qun) He, and Lucas DeBuren

Reviewers

W.R. Everest, W.R. Everest & Associates, Orange, Calif.

E. Kasirga, Fairfax, Va.

S.D.N. Freeman, Black & Veatch, Kansas City, Mo.

N. Voutchkov, Water Globe Consultants, LLC, Winter Springs, Fla.

S. Walker, UTEP Civil Engineering, El Paso, Tex.

The AWWA Technical and Educational Council, the Water Resource Sustainability Division, and the Water Desalting Committee acknowledge the following individuals for their persistence and dedication in developing this manual:

Charlie (Qun) He, *Chair*

Rick G. Bond, *Vice Chair*

J.T. Aguinaldo, Bios Technologies, Tampa, Fla.

B. Alspach, Arcadis, Oceanside, Calif.

S.A. Weikel, AWWA, Denver, Colo.

F. Bloetscher, Public Utility Management & Planning Services Inc., Hollywood, Fla.

R.G. Bond, Black & Veatch, Kansas City, Mo.

Q. Chang, Advanced Engineering and Environmental Services Inc. (AE2S), Fargo, N.D.

L. De-Buren, California State Polytechnic University, Pomona, Calif.

W.R. Everest, W.R. Everest & Associates, Orange, Calif.

S.D.N. Freeman, Black & Veatch, Kansas City, Mo.

V.S. Frenkel, Greeley and Hansen, San Francisco, Calif.

A. Gerling, AWWA, Denver, Colo.

M. Goetz, Hughes & Stuart Communications, Centennial, Colo.

B. Yallaly, Carollo Engineers, Boise, Idaho

C. He, Carollo Engineers, Phoenix, Ariz.

G. Juby, Carollo Engineers, Inc., Riverside, Calif.

E. Kasirga, Fairfax, Va.

M. Mickley, Mickley & Associates LLC, Lafayette, Colo.

T.M. Pankratz, Water Desalination Report, Houston, Tex.

S. Sethi, Carollo Engineers, Sarasota, Fla.

A. Sharbat, California State Polytechnic University, Pomona, Calif.

H. Steiman, Newton, Mass.

D.R. Stewart, Stewart Environmental Consultants, Fort Collins, Colo.

A. Subramani, MWH Americas, Inc., Pasadena, Calif.

S. Veerapaneni, Black & Veatch, Kansas City, Mo.

N. Voutchkov, Water Globe Consultants, LLC, Winter Springs, Fla.

S. Walker, UTEP Civil Engineering, El Paso, Tex.

G.D. Wetterau, CDM Smith, Rancho Cucamonga, Calif.

D. Lesan, Carollo Engineers, Phoenix, Ariz.

Chapter 1

Overview of Inland Desalination

Greg Wetterau and Mike Mickley

According to the United Nations' first edition of the World Water Development Report, 50 percent of the world will face water stress or water shortage by 2025, and 75 percent of the world's population could face water scarcity by 2050 (United Nations 2003). With the world's population increasing and reliable freshwater supplies becoming more limited, desalination has become a growing water supply alternative in many locations. Desalination offers a means of providing freshwater from abundant saline water sources. Its use has historically been most prevalent in coastal areas, treating either seawater directly or seawater-influenced groundwaters. However, as desalination technologies have advanced and the availability of conventional freshwater sources becomes more limited, inland water users have increasingly turned to desalination.

The increased use of inland desalination has created a need for advancements in reliable, cost-effective, and environmentally sensitive means of concentrate management. Coastal desalination facilities often discharge residual concentrate to the ocean; however, with inland facilities, concentrate conveyance lines to coastal areas are often cost prohibitive, requiring alternative concentrate disposal options or the use of costly zero liquid discharge (ZLD) technologies to eliminate concentrate discharge.

This manual presents the key issues and challenges of developing inland desalination facilities, including concentrate management and disposal methods. The manual aims to aid water utilities, industrial facilities, and other water users in evaluating and adapting inland desalination for their water supply needs.

BRACKISH WATER SOURCES

Brackish water can be defined as any water containing between 1,000 and 15,000 mg/L of total dissolved solids (TDS); however, waters with salinity in excess of 15,000 mg/L are sometimes referred to as brackish. Brackish water is present in tidally influenced estuaries, seawater intrusion-impacted groundwaters, inland seas/lakes, and many deep aquifers or shallow aquifers in arid locations. This water was once considered unusable as a drinking water source. However, with the invention and development of reverse osmosis (RO) and other desalination technologies such as electrodialysis, multistage flash distillation, and multiple-effect distillation, brackish waters have become a source for production of drinking water in many arid or water-stressed locations. As this manual is focused on inland desalination, the brackish water sources discussed herein are primarily saline aquifers and inland water bodies rather than tidal or seawater intrusion-influenced waters. While water characteristics of coastal and inland supplies are often similar, the challenges associated with residual and concentrate disposal create unique considerations for an inland brackish water desalination process.

In 2003, the US Geological Survey (USGS) reported that little was known about the hydrogeology of aquifers containing saline water, as groundwater resource evaluations had focused primarily on the prevalence and quality of freshwater aquifers and the freshwater portions of mixed aquifers (USGS 2003). Available evidence indicated that a significant quantity of brackish groundwater exists throughout the United States, although much of this information came from surveys conducted in the 1950s, 1960s, and 1970s. A 2003 study conducted for the Texas Water Development Board found that there is approximately 880 trillion gallons (2.7 billion ac-ft) of brackish groundwater in the aquifers of Texas (LBG-Guyton Associates 2003). In New Mexico, it has been estimated that 75 percent of groundwater is too saline for most uses without treatment.

While it is clear that large quantities of saline groundwater exist throughout the country and that these quantities outnumber freshwater supplies in many water-stressed locations, more study has been needed to better characterize the availability and quality of brackish water in the United States. To work toward that aim, USGS has completed a comprehensive National Brackish Groundwater Assessment (USGS 2017), providing information on brackish groundwater quality and prevalence throughout the mainland United States. Figure 1-1 presents information from this assessment on the depth to brackish groundwater across the country. As part of the assessment, USGS has created a publicly available database containing chemical data and aquifer information for known quantities of brackish groundwater. More recently, the US Bureau of Reclamation and USGS have collaborated on an interactive map tool to increase the ease of use of the National Brackish Groundwater Assessment for water resource planning.

STATUS OF DESALINATION

Global and Domestic Overview

The global use of desalination today exceeds 24 billion gallons per day (bgd) (IDA 2018), employing core technologies developed primarily in the United States with more than $2.0 billion (2018 dollars) of federal research funds between 1959 and 1974 (NRC 2008). With a few exceptions, new desalination plants have predominantly used seawater as their source of supply. Today, 59 percent of the global desalinated water production is from seawater, while only 22 percent is from brackish water, predominantly from groundwater sources. Figure 1-2 shows a breakdown of the source waters for existing desalination facilities. The "ultrapure water" category incorporates various low-salinity sources, including municipal drinking water and local groundwater, to produce high-purity water for industrial use.

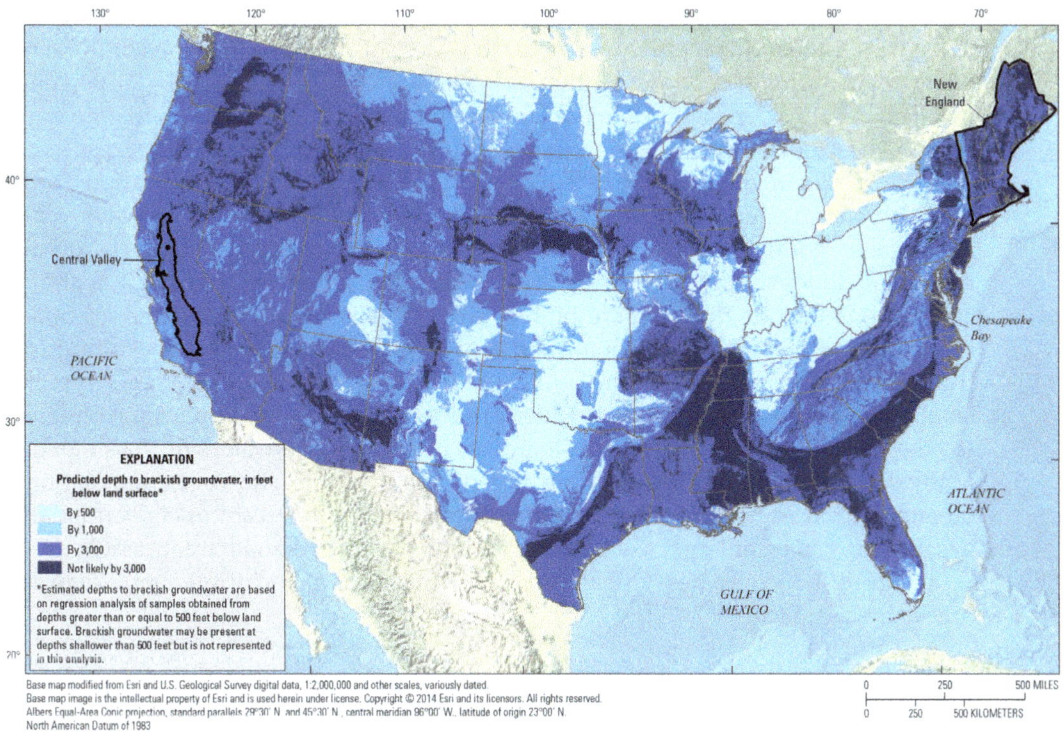

Source: USGS 2017

Figure 1-1 Depth to brackish groundwater in the United States

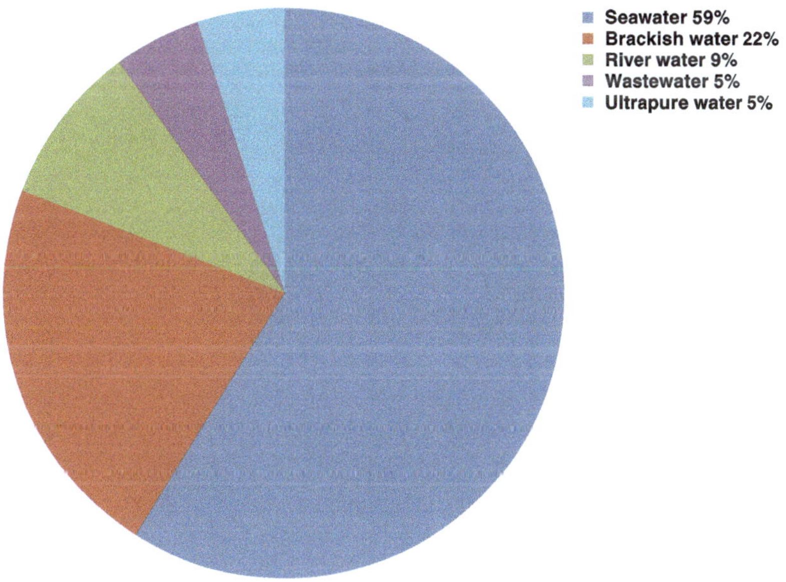

Adapted from International Desalination Association 2018

Figure 1-2 Total worldwide installed desalination capacity by feedwater category

An exception to the global trend is the United States, where the vast majority of installed desalination facilities use brackish water sources. The locations of these brackish water plants remain predominantly in coastal areas; however, nearly 23 percent of the installed desalination facilities in the United States are located in states that do not border any oceans (Mickley 2018).

Most inland desalination is currently used for municipal applications producing potable water, water for disposal or recycle, and facililties producing water for aquifer recharge or aquifer storage and recovery. Nonmunicipal applications of desalination mainly include supply, discharge, and reuse for industry. Industrial water supply applications include those where high-purity water is required, such as for boiler feed, biopharmaceutical manufacturing, food and beverage preparation, and ultrapure water (for semiconductor and pharmaceutical industries). These applications typically are smaller in volume than municipal water production. Discharge applications treat effluent to meet discharge standards, including discharge to surface waters or to sanitary sewers. Many discharge applications are of relatively small volumes; however, applications in the oil and gas industry, including those associated with more recent unconventional oil and gas mining, can be of larger volumes, particularly where regional treatment is used.

Figure 1-3 compares global uses for desalination, including both seawater and brackish water facilities.

US Desalination Facilities

As of 2011, 2.2 bgd (8.5 million m^3/d) of desalination capacity had been installed in the United States. Data on desalination use in the United States are currently limited to municipal facilities, which make up the majority of installed capacity. The municipal facilities use membrane technologies and are overwhelmingly brackish water plants.

A series of detailed surveys of US municipal desalination facilities has been conducted over a 20-year period. The approach of each survey was to identify and contact each plant of size 25,000 gpd (95 m^3/d) or greater built in the 50 US states. The surveys were funded by the Water Research Foundation (WRF), the US Bureau of Reclamation (USBR), St. Johns

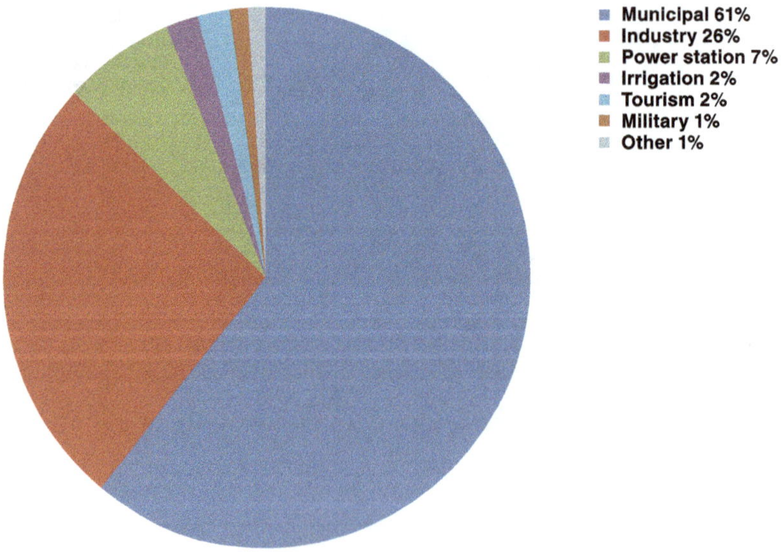

Adapted from International Desalination Association 2018

Figure 1-3 Total worldwide installed desalination capacity by user type

River Water Management District (SJRWMD), and the WateReuse Research Foundation (now the Water Research Foundation). *Development of a Knowledge Base on Desalination Concentrate and Salt Management*, a project published in 2013, provided data for municipal desalination facilities built through 2010. This information was supplemented in 2018 with the publication of the *Updated and Extended Survey of U.S. Municipal Desalination Plants* (Mickley 2018).

Number and types of plants. Data in Figures 1-4 and 1-5 are based on this detailed survey of municipal desalination facilities that provides data for 414 municipal desalination facilities identified from 1971 through 2017 with capacities of 0.025 mgd or greater (Mickley 2018). The number of US municipal desalination plants built with time, and the cumulative capacity of these plants with time, are shown in Figures 1-4 and 1-5,

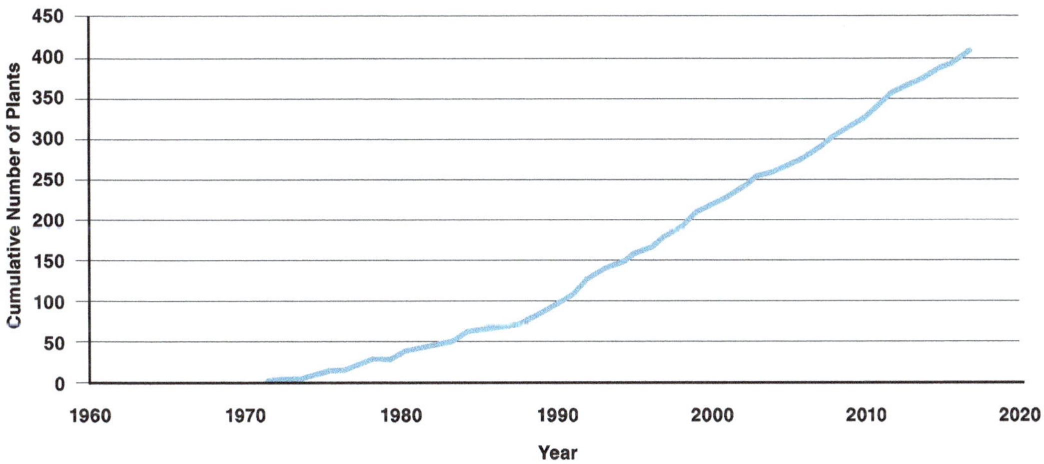

Adapted with permission from Mickley 2019

Figure 1-4 Cumulative number of US municipal desalination plants

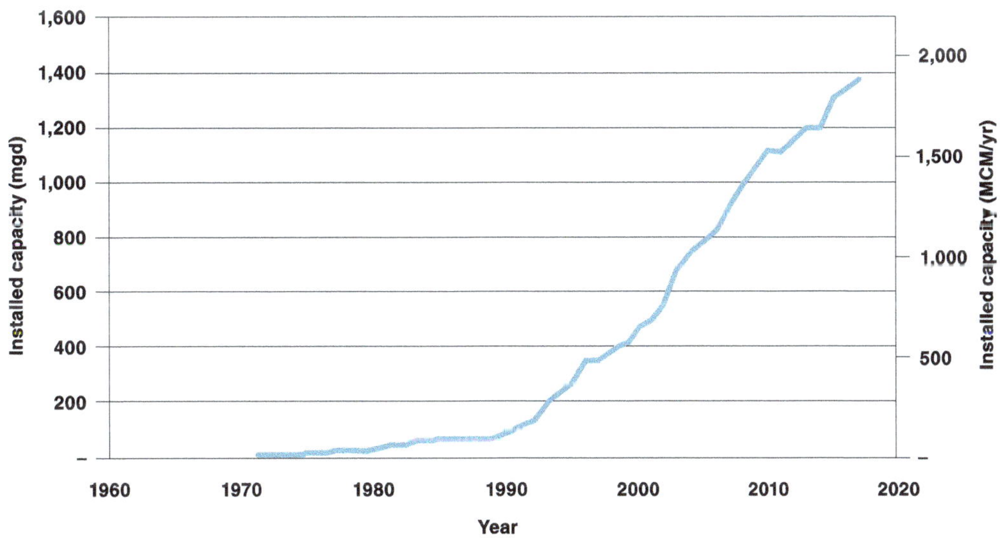

Adapted with permission from Mickley 2019

Figure 1-5 Cumulative installed capacity of US municipal desalination plants

respectively. Capacity is taken as built capacity at the time the plants were surveyed and is presented in both mgd and million cubic meters per year (MCM/yr). The total capacity is an estimate, as many plants have been expanded since they were surveyed and many small, older plants are no longer operating.

Table 1-1 presents the number of plants using various membrane processes at US desalination facilities. It should be noted that these numbers differ slightly from those included in the referenced source (Mickley 2019), as the category "Wastewater RO" used here is different from the integrated membrane system categories used in the source document.

Plant locations. Figure 1-6 shows the existing municipal desalination plants located in 35 states and their distribution. Sixty-seven percent of the facilities (as of 2017) are located in the coastal states of Florida (40 percent), California (14 percent), and Texas (13 percent). The remaining 33 percent are spread over 32 other states, 22 of which have no ocean border.

Table 1-1 Number of different US municipal desalination membrane processes as of 2017

	Number	Percent
Brackish water reverse osmosis (RO)	302	74%
Nanofiltration	61	15%
Electrodialysis reversal	22	5%
Seawater RO	13	3%
Wastewater RO	12	3%

Source: Mickley 2019

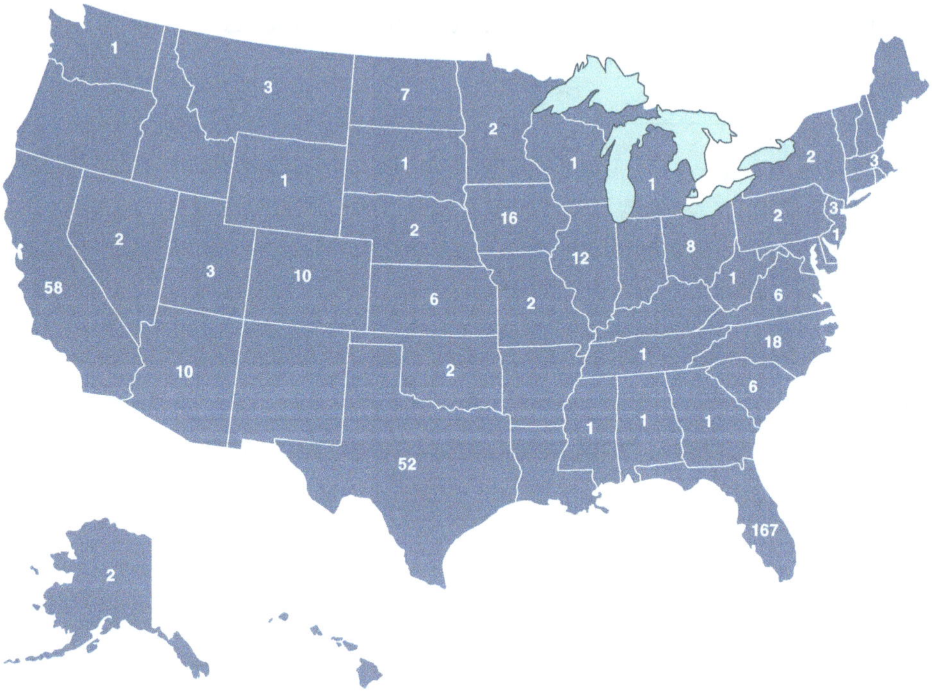

Adapted with permission from Mickley 2019

Figure 1-6 Number of US municipal desalination plants by state as of 2017

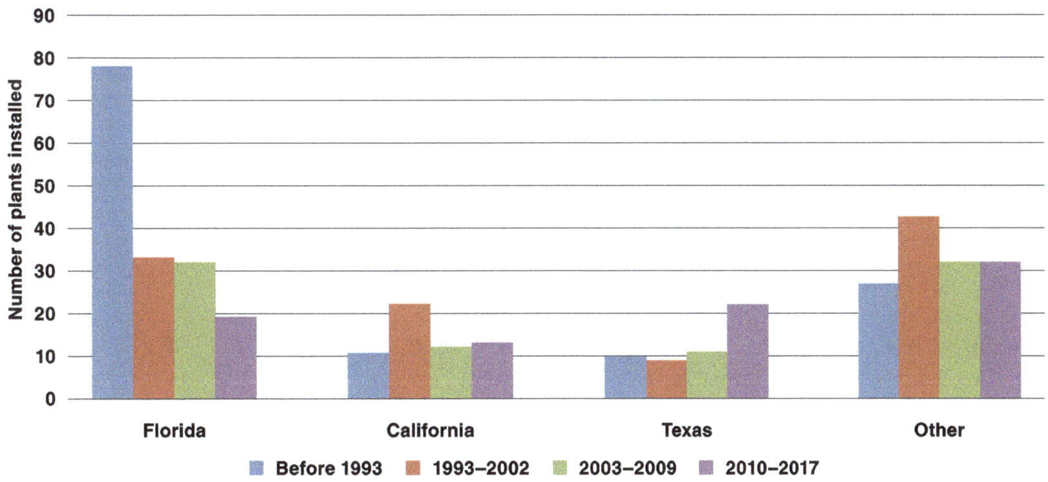

Adapted with permission from Mickley 2019

Figure 1-7 Number of US municipal desalination plants installed by state in different time periods

Figure 1-7 shows the number of plants built in the three major states and the remaining states over four different time periods. Florida was the most active state in all time periods before 2010. During the most recent time period (i.e., after 2010), Texas has become the most active state for new desalination facilities. Since 1993, the percentage of plants being built in Florida has declined from a high of 62 percent in 1993 to the present level of 22 percent. Over this period, the operating capacity of desalination plants has increased from an average of 1.6 mgd per plant built before 1993 to 5.5 mgd per plant built in 2009. During the most recent period, between 2010 and 2017, the average capacity decreased to 3.7 mgd; however, 25 percent of the plants built in this period were in states with no ocean border. With the size of plants increasing, and with more plants being built farther away from coastal areas, the need for reliable concentrate disposal and management alternatives has become more critical.

CONCENTRATE MANAGEMENT OPTIONS

Five conventional concentrate disposal options are currently used by more than 98 percent of the municipal desalination plants built in the United States: surface water discharge, discharge to sewer, deep-well injection, land application, and discharge to evaporation ponds (Mickley et al. 2013). The application of these options has been a function of plant size, water quality, location, regulatory policy, and cost.

Landfill should also be considered as a concentrate disposal alternative for solids produced by high-recovery (HR) processing, where solids result from either accumulation in evaporation ponds or from a final evaporation step to produce mixed solids.

While several possible beneficial uses of concentrate (besides irrigation) have been identified, none are widely applied, most are unproven, and most do not result in the disposal of concentrate. Few viable uses of concentrate have been demonstrated thus far—although some, such as constructed wetlands, may contribute to improved water quality through removal of specific problematic constituents such as selenium or nitrate, making blending and discharge more viable. Given the challenges of concentrate management, it is prudent to explore all beneficial use options early in project planning, as the options are site-specific. A combination of methods linking more conventional options with beneficial uses may provide redundancy, reliability, and potentially some ancillary benefits.

Table 1-2 Concentrate management options

1. Conventional disposal options
   - Surface water discharge
     - Offshore ocean outfall (includes brine line when direct to ocean)
     - Onshore ocean outfall
     - Collocated outfall with coastal power plant or WWTP outfall
     - Discharge to river, canal or lake
   - Discharge to sewer
     - Sewer line
     - Direct line to WWTP
     - Concentrate codisposal with WWTP effluent
     - Hauling of concentrate to a WWTP
   - Subsurface injection
     - Deep-well injection
     - Shallow well (beach well) infiltration
   - Evaporation in ponds
     - Conventional ponds
     - Enhanced evaporation ponds/schemes
   - Land application
     - Percolation ponds/rapid infiltration basins
     - Irrigation
2. Landfills (for solids)
   - Dedicated monofills
   - Industrial landfills
3. Beneficial use (other than irrigation, which was addressed previously under "land application")

Source: Mickley 2019

The use of the term *concentrate management* (rather than *concentrate disposal*) and the use of higher-recovery processing are relatively new to municipal desalination. As a result, some publications list ZLD and near-zero liquid discharge (NZLD) as concentrate management options. This view is correct from the perspective of having an existing concentrate and considering options for how to manage it. However, in most other industries, ZLD and NZLD are processing options that are considered and possibly implemented with the initial design of the treatment system. Table 1-2 presents concentrate management options from this perspective, including all options for the final fate of concentrate, or solids, whether from a low- or high-recovery process.

Table 1-3 presents the percentage use of the five conventional disposal options and the number of states having municipal desalination plants using the option. Thirty-five states were identified as having desalination plants practicing the concentrate management options listed in Table 1-2. Few of these states use deep-well injection, land application, or evaporation ponds. The states using disposal alternatives other than surface discharge or discharge to sewers or wastewater treatment plants (WWTPs) are presented in Table 1-4. In summary:

- Seventy percent of the plants discharge concentrate to surface water, to the sewer, or to a WWTP.
- Deep-well injection and land application are used in only five of the 35 states.
- Evaporation ponds are used in only four of the 35 states.

- In 28 states, 100 percent of the plants discharge concentrate either to a surface water body or to the sewer.
- In 2017, all but six of the 70 deep-well injection sites were in Florida.
- Twenty of the 27 land application sites are in Florida.
- Florida, California, and Texas are the only states using all five conventional disposal options.

The pattern of use of concentrate disposal options varies greatly in the three states with the largest number of desalination plants (Florida, California, and Texas). These are three of the four most populous states and have experienced the greatest growth in residents since 2000 (Mickley et al. 2013). They have thus experienced the greatest increase in potable water demand. These three coastal states are also the centers of interest and activity for large-scale seawater RO plants. Florida, California, and Texas may be considered as distinct regions because of the number of plants, the resulting awareness of the regulatory agencies and the public to desalination and concentrate management issues, and the different patterns of use of the five conventional disposal options. In particular, the following conclusions can be drawn about the plants in each region:

- California plants primarily discharge to surface water (ocean) and dispose to sewer.
- Texas plants primarily discharge to surface waters and evaporation ponds.
- Florida plants use a mix of methods with the highest percentage of use of deep-well injection.

The other 32 states that have municipal desalination plants rely almost exclusively on discharge to surface water and disposal to sewer.

While those conclusions represent the trends of facilities built to date, the trend toward larger desalination facilities in more inland locations should be expected to have an impact on the types of concentrate disposal options used. As plants grow larger, the impacts of concentrate disposal on the surface water body or watershed become more significant and will push more facilities toward alternative concentrate disposal options.

Table 1-3 Number of states using the disposal options for municipal desalination concentrate

	Percentage use	Number of states
Surface water discharge	46	27
Discharge to sewer or wastewater treatment plant	24	24
Deep-well injection	17	4
Land application	7	4
Evaporation in ponds	5	3

Source: Mickley 2019

Table 1-4 States using various disposal options and number of plants in each state

	FL	CA	TX	KS	AZ	PA	CO
Deep-well injection	64	1	2	1	0	0	2
Land application	20	1	5	0	1	0	0
Evaporation ponds	3	2	13	0	3	0	0
Recycle	0	4	4	0	1	1	0

Source: Mickley 2018

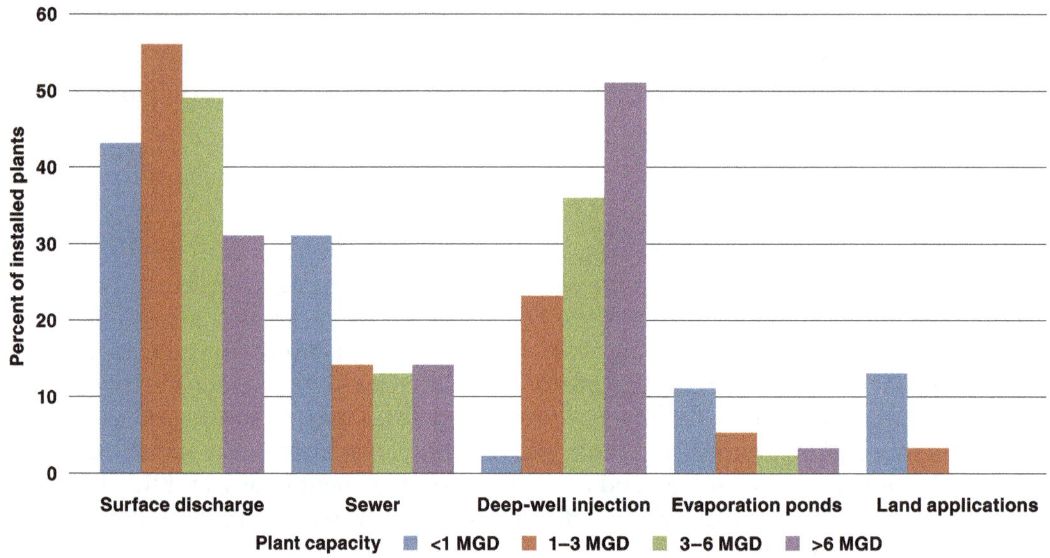

Adapted with permission from Mickley 2019

Figure 1-8 Concentrate disposal options for US municipal desalination plants

Figure 1-8 shows a breakdown of concentrate disposal alternatives by plant capacity. Based on the trends present in these numbers, and on a comparison of plants built before 1993 with those built after 2002, the following conclusions can be drawn:

- Relative to plants built before 2002, there has been greater use of deep-well injection and a declining use of evaporation ponds or land application.
- In the past several years, there has been increased consideration and investigation of deep-well injection in states other than Florida as well as enhanced evaporation. However, neither has had much impact on practices.
- An increasing number of plants are treating source water to remove specific contaminants as well as to reduce salinity.
- An increasing number of plants have concentrate containing contaminants that restrict application of concentrate management options or require treatment to remove the contaminants before disposal.
- Increasing concentrate management challenges have led to planning-phase consideration of plants of HR processing of concentrate. To date, a few ZLD plants have been built (Mickley 2018), and a few HR nanofiltration plants are now being built.

Trends in US Concentrate Management

Although "management" is the broader and more inclusive term for addressing handling of concentrate flows, nearly all concentrate from desalination plants is currently discharged to surface water bodies or subsurface formations. The primary challenge of inland concentrate management continues to be finding a cost-effective and environmentally sustainable option (NRC 2008).

The time and effort required to find a solution, in large part defined by the regulatory framework and permitting process, are increasing in most locations. An emerging issue is that more stringent regulations for protecting source water will pose additional impediments to discharge options that can affect source water. These include surface water discharge and disposal to sewer, which together currently are used by 80 percent of the municipal desalination facilities. An increasing number of water reclamation plants are treating source water to remove contaminants and reduce salinity. As a result, as was found in the most recent municipal desalination concentrate management survey, a handful of desalination plants have concentrate containing contaminants that restrict discharge of concentrate or require treatment to remove the contaminants before disposal (Mickley et al. 2013).

This trend is likely to continue as regulations become more stringent because of nutrients, emerging contaminants, and other contaminants being considered for regulation. Because of these and other challenges, there has been increased interest in HR processing, usually under the label of volume reduction or concentrate minimization. This has been reflected in funded research projects and more frequent consideration of HR processing at the planning phase of plants. Several plants now incorporate HR processing, and a few ZLD plants are operational, treating drinking water desalination concentrate in the United States (Mickley et al. 2018).

MANUAL OVERVIEW

This manual contains the following chapters:

Chapter 2, *Water Quality and Planning Strategies*, defines the water quality challenges and goals associated with inland desalination and the general planning strategies involved in developing a brackish water supply and managing concentrate.

Chapter 3, *Brackish Water Desalination*, describes the most commonly used technologies involved with brackish water desalination, detailing the key issues associated with facility design and the primary treatment components of a brackish water desalination facility.

Chapter 4, *Discharge Options for Concentrate Disposal*, describes concentrate management options that do not involve treatment of concentrate to recover part or all of it as product water.

Chapter 5, *Enhanced Recovery and Zero Liquid Discharge*, discusses technologies used to support zero liquid discharge or near-zero liquid discharge desalination facilities, including benefits and challenges of various process trains currently in use or in development.

Chapter 6, *Cost of Desalination and Concentrate Management*, provides a general discussion of the cost of inland desalination, including concentrate disposal, and the primary factors impacting the associated costs.

Chapter 7, *Regulatory, Safety, Operational, and Environmental Issues*, discusses key regulatory challenges with implementing inland desalination as well as safety, operational, and environmental concerns that must be addressed when planning for and developing a new facility.

Chapter 8, *Case Studies*, includes detailed information on a wide variety of inland desalination projects, including lessons learned from facilities already using brackish supplies to address their water needs.

Chapter 9, *Technological Trends, Salt Recovery, and Beneficial Uses*, discusses technology advancements on emerging concentrate treatment alternatives, including salt recovery, nutrient recovery, and beneficial use of concentrate.

REFERENCES

IDA (International Desalination Association). 2018. *Worldwide Desalting Inventory*. Vol. 30. Topsfield, Mass.: IDA

LBG-Guyton Associates. 2003. *Brackish Groundwater Manual for Texas Regional Water Planning Groups*. Austin, Tex.: LBG-Guyton Associates.

Mickley, M. 2019. *Updated and Extended Survey of U.S. Municipal Desalination Plants*. DWPR Program Report No. 207. Denver: US Bureau of Reclamation.

Mickley, M. 2018. *Updated and Extended Survey of U.S. Municipal Desalination Plants*. Denver: US Bureau of Reclamation.

Mickley, M.; Jordahl, J.; and Arakel, A. 2013. *Development of a Knowledge Base on Desalination Concentrate and Salt Management*. Alexandria, Va.: WateReuse.

Mickley, M. 2012. US Municipal Desalination Plants: Number, Types, Locations, Sizes, and Concentrate Management Practices. *IDA Journal of Desalination and Water Reuse*, 4(1):44.

NRC (National Research Council). 2008. *Desalination: A National Perspective*. Washington: The National Academies Press.

United Nations. 2003 (1st ed.). World Water Development Report. Paris: United Nations Educational, Scientific and Cultural Organization.

USGS (United States Geological Survey). 2017. *Brackish Groundwater in the United States*. https://pubs.usgs.gov/pp/1833/pp1833.pdf (accessed Nov. 30, 2018).

USGS. 2003. *Desalination of Ground Water: Earth Science Perspective*. Reston, Va.: USGS.

Chapter 2

Water Quality and Planning Strategies

Charlie (Qun) He, Arun Subramani, Greg Wetterau, and Vasu Veerapaneni

This chapter provides an overview of planning strategies for implementing inland desalination and concentrate management. The need for desalination and concentrate management may be driven by different factors. For example, concentrate may be generated by the desalination process when an additional water supply is developed to meet increasing demands from new sources of impaired water quality. Moreover, a source water with better water quality [e.g., lower concentrations of total dissolved solids (TDS) and target ions] is desired for special agricultural or industrial needs. This chapter focuses on planning driven by water scarcity challenges.

Desalination and water reuse are potential solutions to global water scarcity issues; however, concentrate management is an unavoidable result of these solutions. This is because water reuse often relies on desalination technologies such as reverse osmosis (RO) to achieve its removal goals for dissolved solids, emerging contaminants, and pathogens. However, after water is used for many cycles, dissolved salt will accumulate and interfere with the reuse efficiency to an extent that the water has to be desalinated for any further use. In both cases, once desalination technologies are employed, a concentrate waste stream is generated, requiring proper management and disposal.

Concentrate disposal options vary from one system to another, and several concentrate disposal options should be considered before employing complex concentrate treatment and management technologies. Sewer discharge may be the simplest and cheapest option, but it is often not sustainable as the salt content in the reclaimed water increases. For most inland communities, surface water discharge to an ocean or a salt lake nearby is not feasible. Deep-well injection could be a solution, but viable hydrogeological conditions may not exist for every inland application and the injection could negatively impact seismic conditions. Even when they are available, the capacity of confined aquifers suitable for deep-well injection may be limited, requiring minimizing the volume of the concentrate.

For most areas with little precipitation, evaporation ponds remain a reliable resource. Evaporation ponds are relatively easy to construct, require little maintenance, and can be expanded as needed to avoid significant investment all at once. However, for a brackish water RO system at municipal scale with a typical recovery of 70–85 percent, evaporation ponds without concentrate volume reduction could occupy square miles of land and potentially result in other environmental issues. Some enhancement options (e.g., mechanical mixing and wind aided) could be applied to reduce evaporation pond size.

If concentrate disposal options are limited by available land for evaporation ponds, treating concentrate to reduce its volume and removing limiting constituents that prevent high recovery may become necessary. Concentrate treatment is often more difficult than conventional water treatment and handling of solid phase residuals. Concentrate management is complex; it is affected by many factors and involves multiple steps. Generally, the optimal approach to concentrate management involves a combination of several technologies and strategies.

The quality of natural water sources can vary significantly from one location to another. In addition, water quality variations are amplified in the brine streams after being concentrated. Currently, there is no single technology or methodology that can resolve all salinity management challenges for all situations. The best approach often lies in developing site-specific solutions. The process of developing site-specific solutions to salinity management challenges often involves the following steps:

- Evaluate brackish source water quality and quantity as well as the characteristics of the contaminants limiting concentrate management options.
- Establish water quality goals for the site-specific conditions of a given project.
- Identify treatment strategies based on water quality goals and limiting factors, and provide a basis for planning strategies.
- Combine strategies and solutions into a cost-effective project design.
- Demonstrate efficiencies of critical components through bench-scale and pilot testing.
- Optimize and enhance customized strategies and solutions.

These steps are important in ensuring that water quality and recovery goals are met while minimizing life-cycle costs.

SOURCE WATER QUALITY AFFECTING DESALINATION

This section discusses how brackish source water quality affects desalination and treatment goals; water quality monitoring recommendations; guidance on desalination technology performance projection models; an overview of various treatment levels and desalination and concentrate management technologies; and key planning strategies that apply to inland water utilities developing a water supply from brackish sources.

Brackish Groundwaters

Brackish groundwaters vary considerably by region and often by depth in similar locations. Table 2-1 presents an overview of the water quality of five characteristic brackish water sources currently used for desalination supplies.

Table 2-1 Example of brackish water quality variations

Parameter	South Florida[1]	West Texas[2]	Central Colorado[3]	Southern California[4]	Central Arizona[5]
TDS (mg/L)	3,680	1,125	847	1,700	2,283
pH	7.6	8.2	7.2	7.4	7.5
Calcium (mg/L)	117	67	145	129	188
Magnesium (mg/L)	143	16	40	53	77
Sodium (mg/L)	1,070	295	122	405	606
Potassium (mg/L)	64	12	3.7	7	6
Barium (mg/L)	0.01	0.2	0.05	0.08	0.09
Strontium (mg/L)	11.2	1.5	ND	0.9	ND
Bicarbonate (mg/L)	145	124	308	239	362
Sulfate (mg/L)	440	71	192	181	436
Chloride (mg/L)	2,240	504	136	760	914
Fluoride (mg/L)	ND	0.7	1.5	0.13	1.1
Nitrate-N (mg/L)	ND	3.1	2.6	0.07	18
Silica (mg/L)	8.6	31	8.0	17	27
Iron (mg/L)	ND	ND	0.06	0.09	ND
Manganese (mg/L)	ND	ND	0.07	0.6	ND
Sulfide (mg/L)	2.4	ND	ND	ND	ND
Aluminum (mg/L)	ND	ND	0.008	0.05	ND

ND—No data, TDS—total dissolved solids
[1] South Florida aquifer, Deerfield Beach, Fla.
[2] Surficial aquifer, El Paso, Tex.
[3] Surficial aquifer, East Cherry Creek, Colo.
[4] San Diego Formation, Chula Vista, Calif.
[5] Buckeye and Goodyear, Ariz.

The South Florida and West Texas waters are examples of sodium- and chloride-dominated sources in which sodium and chloride make up most of the TDS concentration. In the case of the South Florida water, which has low concentrations of both calcium and sulfate, 90 percent of the source water is sodium and chloride. The Central Colorado water is an example of a source that contains almost equal concentrations of sodium, chloride, calcium, and sulfate. The Southern California water source is also mostly sodium and chloride but has moderate calcium, sulfate, and bicarbonate. West Texas and Central Arizona water has a significantly higher concentration of silica compared to the other sources, which would be a factor in selecting an appropriate desalting technology and recovery rate.

Brackish water quality parameters affecting desalination. The desalination process is directly affected by source water quality constituents such as TDS—organic and inorganic—and origin (e.g., groundwater, surface water, wastewater effluent). The cost of treating brackish water highly depends on the water quality of the source. Waters with higher salinity, higher sparingly soluble salts, or higher oxidizable metals will typically be more costly to treat than water supplies with lower concentrations of these dissolved constituents. The presence of hydrocarbons can also create challenges to using brackish groundwaters.

The following sections discuss the impact that suspended solids, inorganic compounds, minerals, and oxidizable metals have on the desalination process. The treatment processes most commonly used to remove contaminants from brackish water, including RO, nanofiltration (NF), electrodialysis (ED), and electrodialysis reversal (EDR), are discussed in Chapter 3 of this manual. AWWA M46, *Reverse Osmosis and Nanofiltration* (AWWA 2007), and AWWA M38, *Electrodialysis and Electrodialysis Reversal* (AWWA 1995), also offer more detailed discussions of the treatment technologies used to remove these constituents.

Suspended solids. RO and NF membranes effectively remove not only dissolved constituents but also particulate and colloidal materials. However, the latter is a misapplication of the technology. Suspended solids can accumulate on the RO membrane surface and clog feed channels, resulting in poor performance and sometimes irreversible damage to the membranes in a relatively short time. Even in small concentrations, suspended solids in the source water that accumulate on the membrane surface deposit a foulant layer that reduces permeability and production. The foulant deposits may also increase the thickness of the concentration polarization layer at the membrane surface, resulting in increased salt passage and higher energy consumption. Hence, if solid or colloidal materials are present in the water to be treated, the RO or NF process should be preceded by pretreatment. Essentially all facilities include at least a cartridge filtration step before RO or NF to protect the membrane elements.

One of the most widely used indicators for determining the suspended solids and colloidal fouling potential of desalination feedwater is the measurement of silt density index (SDI). Typically, the target $SDI_{15\text{-min}}$ for RO and NF feedwater is less than 3–5. Unlike pressurized membrane processes, ED and EDR do not remove suspended solids and can tolerate a higher concentration of feedwater suspended solids. Manufacturers recommend a target feedwater $SDI_{5\text{-min}}$ of less than 10–15 for EDR.

Inorganic compounds. Both RO and EDR can effectively remove a wide range of inorganic contaminants (Antony et al. 2011). As a result, they could both be affected by scaling of inorganic compounds. With the exception of silica, the extent of such impacts is very similar for RO and EDR. The following discusses impacts of oxidizable metals and minerals, sparingly soluble salts, and silica on desalination processes.

Oxidizable metals and minerals. The presence of iron, manganese, or hydrogen sulfide in a brackish water supply can be problematic for desalination facilities, as these ions are easily oxidized by ambient oxygen and can cause excessive particulate fouling when oxidized. Iron and manganese do not present a fouling risk when present in their reduced state, and desalination facilities have successfully operated with iron and manganese levels exceeding 2–3 mg/L, provided anoxic conditions are maintained within the wells, transmission piping, and treatment facility. Reduced and oxidized forms of iron and manganese are removed by RO, NF, and EDR membranes and therefore do not create the risk of red or black water in desalinated product water. However, if iron and manganese are already in oxidized forms, $Fe(OH)_3$ and $Mn(OH)_3$ may precipitate out and foul the membrane.

If iron and manganese cannot be conveyed to the desalination process without exposure to oxygen in the air or other oxidizing agents, pretreatment is necessary. Typically, if iron and manganese concentrations are greater than 0.05 mg/L in the feedwater and anoxic conditions cannot be maintained to prevent them from being oxidized, pretreatment using oxidation and filtration is recommended to remove iron and manganese before desalination. Numerous facilities have experienced rapid and repeated plugging of membrane elements when oxidized iron and manganese have not been adequately removed. The presence of iron exacerbates silica fouling, which is discussed later in this chapter.

Similar to iron and manganese, an insoluble form of aluminum can form $Al(OH)_3$ in the desalination system. Aluminum can be introduced to the feedwater through the use of coagulants in pretreatment processes and its carryover to the desalination process. Aluminum solubility is lowest around pH 6.5, and the pH suppression commonly used to control fouling from other sparingly soluble salts can therefore make aluminum fouling worse. When aluminum fouling is present, alternative coagulants can be considered.

Hydrogen sulfide, similar to iron and manganese, does not present a fouling risk when anoxic conditions are maintained. However, hydrogen sulfide is not well rejected by RO or NF membranes and often must be removed from the desalination product water through degasification. Where anoxic conditions cannot be maintained, hydrogen sulfide may be oxidized to elemental sulfur, causing excessive fouling on desalination membranes.

Sparingly soluble salts. Sparingly soluble salts are those minerals that have a lower solubility, such as calcium sulfate, and can create complications in the desalination of brackish water. As the membrane recovers permeate from the feedwater, the concentrations of some of the sparingly soluble salts and silica can exceed solubility thresholds and form a precipitate (scale) on the membranes. Scale formation leads to a decrease in permeability, deteriorating the quality of water produced and requiring an increase in operating pressure and frequent chemical cleanings. The common precipitates that limit desalination recovery are calcium carbonate, calcium sulfate, calcium fluoride, calcium phosphate, barium sulfate, strontium sulfate, and silica. Usually only one or two of these constituents are limiting in each situation. Identifying and addressing the limiting factor are critical for planning and designing desalination facilities.

Acid addition and polymeric antiscalants are used to reduce the risk of scaling on the membranes, with different antiscalants tailored for different types of scaling agents. Commonly used RO design software will calculate the solubility of many sparingly soluble salts, such as calcium carbonate, calcium sulfate, strontium sulfate, barium sulfate, and silica. However, some desalination facilities have reported challenges with other salts, such as calcium phosphate, calcium fluoride, and aluminum silicate, which are less commonly modeled in most vendor software. Desalination facilities struggling with scale from sparingly soluble salts will often perform membrane "autopsies" to identify specific scaling agents and adjust operating conditions or cleaning approaches to reduce the risk of scaling. Scale can be removed through acid cleaning of the membranes; however, excessively scaled membranes can become damaged or permanently fouled, requiring membrane replacement.

Silica. Silica is generally found in three forms in natural waters—namely, reactive silica (dissolved silica), colloidal silica, and particulate silica. Reactive silica in water can be found in many forms, e.g., monosilicic acid, disilicic acid, and polysilicic acid with molecules smaller than the colloidal range (Ning et al. 2006). Reactive silica, $Si(OH)_4$, is monomeric and contains a single silicon atom. The dissolution rate depends mainly on pH and temperature.

Colloidal silica is widely thought to be both silica that has polymerized into silicon dioxide and silica that has formed complex compounds with organic and/or inorganic substances in the source water. Particulate silica is larger in size and mostly consists of sand or suspended solids. Silica polymerization on the membrane leads to increased chemical cleaning frequency and may eventually lead to replacement of membrane elements. The presence of silica can be problematic for pressure-driven desalination processes (NF and RO), but is less of a concern for electrochemical-driven processes (ED and EDR), which do not concentrate or remove silica.

For reducing silica polymerization on the membrane surface, operation at either low pH (<5.0) or high pH (>10.0) is desirable. The use of antiscalant specific to silica is also beneficial for restricting silica fouling on the membrane surface. Chemicals used for silica fouling control can be divided into three categories:

- chemicals for inhibition of silica scale formation (silica inhibitors),
- chemicals for silica dispersion, and
- chemicals for silica scale removal from membrane surfaces.

Silica inhibitors can retard the polymerization of silica at supersaturated concentrations. Dispersants place a surface charge on the polymerized silica and cause repulsion and dispersion of these polymerized structures to prevent formation of colloidal silica (Sheikholeslami and Tan 1999). Besides using antiscalant, removing silica through a process such as lime softening is effective and is discussed in Chapters 5 and 8.

In the western states, the recovery rates of RO systems often may be limited by the solubility of silica. When silica is present at high concentrations, precipitation and deposition of amorphous silica and silicates could result in consequential RO membrane fouling and loss of efficiency. Particulate and colloidal silica cause severe, irreversible fouling that is difficult to remove.

Organics and Hydrocarbons

Desalination membranes can be fouled by organic constituents, depending on the interactions between the natural organic matter in the feedwater and the RO membrane. This is typically an issue in surface water sources such as river water, seawater, and wastewater. Less commonly, brackish groundwater sources contain high concentrations of organic material, except in regions such as Florida. If present, organic material can foul the RO membrane, interfere with the functioning of the antiscalant, and potentially aggravate biofouling. Typically, organic fouling is reduced by optimizing pretreatment, adjusting the operating pH, and using compatible antiscalants.

Biofouling is one of the most challenging fouling phenomena that is prevalent in surface water and wastewater sources. Biofouling commonly occurs when the feedwater has a high concentration of organic material and nutrients. The large area of membranes provides surface areas for microbial growth. Typically, colonization of the microorganisms results in the formation of biofilm on the surface of the membrane because of the production of extracellular polymers that attach to the membranes. Other foulants typically get imbedded within this biofilm, increasing the severity of the fouling. Biofouling is controlled by adequate pretreatment of feedwater, reduction of stagnant feedwater in the membranes when not in service, use of oxidants that do not damage the membranes, and periodic chemical cleaning of membranes. In reuse applications treating wastewater containing ammonia, chlorine is added in pretreatment to form chloramines, which continue through the desalination system to help minimize biofouling.

Hydrocarbons can cause severe and sometimes permanent fouling of desalination membranes, adsorbing to the membrane surface and restricting water flow. RO vendors typically limit oil and grease in membrane feedwater to 0.1 mg/L or lower to prevent such fouling. RO facilities treating water high in hydrocarbons have been successful at reducing fouling by operating under high pH conditions or by implementing significant levels of pretreatment to reduce residual oils and greases.

PRODUCT WATER QUALITY GOALS

Product water quality goals vary depending on the intended use of the water. This section provides a brief overview of water quality requirements for various uses. Some utilities may have goals that differ from those listed in this section.

Potable Water Quality Goals

In the United States, the quality of potable or drinking water is regulated by the Safe Drinking Water Act (SDWA), including the National Primary Drinking Water Regulations (NPDWR) and the National Secondary Drinking Water Regulations (NSDWR). A detailed list of these standards can be found on the US Environmental Protection Agency (USEPA) website. Other regions of the world have similar drinking water standards (Guidelines for Canadian Drinking Water Quality, European Commission Standards, etc.). Most developing countries follow the World Health Organization guidelines for drinking water.

Table 2-2 provides a list of typical water quality goals for various inorganic contaminants; this list can be used as a reference when developing water sources. Some of these goals are based on drinking water quality standards, noted with their maximum contaminant level (MCL) or secondary MCL (SMCL). Many utilities or local regulatory agencies may choose to use a percentage of the MCL and SMCL levels (e.g., 80 percent) as their targets to provide additional operational buffer. Many factors such as TDS and temperature affect consumer taste preferences (Gallagher and Dietrich 2010). Some utilities successfully conducted outreach campaigns such as Decision H_2O by the City of Tucson, Ariz., to involve the public in determining an acceptable and affordable mineral content in the drinking water (Gallagher and Dietrich 2010).

Water that is high in calcium and magnesium is often referred to as "hard water." While hardness is not a health risk and is not regulated, it causes unpleasant aesthetic

Table 2-2 Examples of drinking water quality goals

Contaminant	Primary MCL or TT* (mg/L)	Secondary MCL (mg/L)
Antimony	0.006	
Arsenic	0.010 as of 01/23/06	
Barium	2	
Beryllium	0.004	
Cadmium	0.005	
Chromium (total)	0.1	
Copper	TT†; Action level=1.3	1.0
Cyanide (as free cyanide)	0.2	
Fluoride	4.0	2.0
Lead	TT†; Action level=0.015	
Mercury (inorganic)	0.002	
Nitrate (measured as nitrogen)	10	
Nitrite (measured as nitrogen)	1	
Selenium	0.05	
Thallium	0.002	
Aluminum		0.05–0.2
Chloride		250
Iron		0.3
Manganese		0.05
Silver		0.1
Sulfate		250
Total dissolved solids (TDS)		500
Zinc		5

* TT – treatment technique; a required process intended to reduce the level of a contaminant in drinking water.

† Copper and lead are regulated by a treatment technique that requires systems to control the corrosiveness of their water. If more than 10% of tap water samples exceed the action level, water systems must take additional steps. For copper, the action level is 1.3 mg/L; for lead, 0.015 mg/L.

Table 2-3 Examples of short-term salinity goals

	Maximum Contaminant Level Ranges		
Constituent, Units	Recommended	Upper	Short Term
Total dissolved solids (TDS), mg/L*	500	1,000	1,500
Specific conductance, μS/cm*	900	1,600	2,200
Chloride, mg/L	250	500	600
Sulfate, mg/L	250	500	600

Source: Barclays Official California Code of Regulations, Title 22. Social Security, Division 4. Environmental Health, Chapter 15. Domestic Water Quality and Monitoring Regulations, Article 16. Secondary Drinking Water Standards. §64449 Secondary Maximum Contaminant Levels and Compliance.

* Meet TDS or specific conductance goals.

impact on laundry, bathing, and water heaters. Water fixtures and pipes can become clogged with calcium and magnesium scale that reduces flow and requires their replacement. Softeners are often used to remove hardness, but they do not remove TDS.

Desalination technologies such as RO and ED are effective for removal of chloride, hardness, TDS, and other inorganic substances. However, these technologies, especially when considering the associated concentrate management, are fairly expensive and may be impractical. Some inland utilities and regulatory agencies may choose to establish a target range or temporary goals similar to the values presented in Table 2-3.

Nonpotable Reuse

When reclaimed water is reused for nonpotable reuse (e.g., for irrigation), several issues could result from high concentrations of salt and certain ions individually or in combination. The degrees of impact depend on time, concentration, crop sensitivity, and crop water use.

High-TDS water limits the ability of certain crops and grasses to grow and flourish. Salt buildup in the root zone is a prevalent problem. Flushing these salts involves applying more water than normal, dissolving the existing salts in the plant's root zone, and "flushing" those salts further into the ground. This increases the water usage and reduces the irrigation efficiency.

A critical constituent of TDS that requires specific attention is sodium ion (Na^+). High sodium concentrations in soil result in poor infiltration. Sodium adsorption ratio (SAR) is a measure of the suitability of water for use in agricultural irrigation. The formula for calculating sodium adsorption ratio is as follows:

$$SAR = \frac{Na^+}{\sqrt{\frac{1}{2}\left(Ca^{2+} + Mg^{2+}\right)}} \qquad \text{(Eq 2-1)}$$

where

Na^+ = sodium ion concentration

Ca^{2+} = calcium ion concentration

Mg^{2+} = magnesium ion concentration

All concentrations are expressed in milliequivalents/liter (meql/L).

According to the US Bureau of Reclamation (USBR), water with SAR less than 3 is safe for irrigating turf and ornamental landscape plants (USBR 2003a, 2003b). SAR greater than 9 can cause severe damage to the soil condition. However, SAR is only one of the factors that determine the suitability of water for irrigation (Clark et al. 2000, Maas 1984). For example, if irrigation water contains relatively high amounts of bicarbonate ion, the bicarbonate can affect the calcium and magnesium concentrations in the soil to which the water is applied. For this reason, the water's bicarbonate and total salinity are often taken into account (in addition to its calcium, sodium, and magnesium concentrations) and reported as adjusted SAR. In general, the higher the SAR, the less suitable the water is for irrigation. Some utilities have adopted treatment goals for sodium (e.g., sodium < 125 mg/L; Chesters et al. 2007) and SAR goals (e.g., SAR < 5) for their reclaimed water reuse facilities.

Ayers and Westcot (1985) gave the known tolerances of several crops to sodium, chloride, and boron. These values may be used as a reference when establishing water quality goals, supplemented by local experience and comprehensive environmental impact assessments.

Potable Reuse

For potable reuse, reclaimed water must be treated to meet drinking water regulations similar to those being developed by the California State Water Resource Control Board Division of Drinking Water (DDW) and the Texas Commission on Environmental Quality (TCEQ). The World Health Organization (WHO 2017) provided information on specific aspects of potable reuse, including the water quality, types of treatment processes, additional monitoring considerations, potential use of environmental buffers and engineered storages, and public acceptance. WateReuse Research Foundation (WRRF), AWWA, Water Environmental Federation, and National Water Research Institute (NWRI) provided a detailed framework for the implementation of direct potable reuse (DPR), covering not only water quality–related public health standards or quality measures but also means to ensure that such standards can be reliably met (WRRF 2015). The USEPA has also developed guidance on potable reuse through the 2017 Potable Reuse Compendium (USEPA 2017). Examples of water quality criteria for potable reuse are provided in Tables 2-4 and 2-5.

To meet a 500 mg/L TDS regulatory limit, a desalination process such as RO or NF is often required. However, as part of a potable reuse train, RO may be used for more than just desalination to remove other chemical, microbial, or radiological contaminants. When RO is used, a concentrate stream will be generated and must be handled appropriately. In areas where concentrate disposal options are limited, consider whether a non-RO–based potable reuse scheme could be sufficient to avoid concentrate management.

WATER QUALITY ANALYSIS AND MONITORING

Accurate and representative water quality data provide a sound basis for planning desalination and concentrate management facilities. Furthermore, for a facility in operation, water quality monitoring continues to provide critical information for process control and operational optimization.

Table 2-4 Regulations in California for groundwater recharge into potable aquifers

Minimum Criteria
>12-log virus inactivation >10-log *Giardia* and *Cryptosporidium* inactivation
<2.2. total coliform/100 mL
<0.5 mg/L TOC for 100% recycled water contribution <5 mg/L total nitrogen
Drinking water MCLs

MCLs—maximum contaminant levels; TOC—total organic carbon

Table 2-5 Example of product water quality criteria for direct potable reuse applications

Microbial Group	Criterion
Enteric virus	12 log removal
Cryptosporidium spp.	10 log removal
Total coliform bacteria	9 log removal
Chemicals	**Criterion**
Trihalomethanes	80 μg/L
Halogenated acetic acids	60 μg/L
N-Nitrosodimethylamine	10 ng/L
Bromate	10 μg/L
Chlorate	800 μg/L
Perfluorooctanoic acid	0.4 μg/L
Perfluorooctane sulfonate	0.2 μg/L
Perchlorate	15 μg/L
1,4-Dioxane	1 μg/L
Steroid Hormones	**Criterion**
Ethynyl Estradiol	None. Will be in ng/L level if established.
17-b-Estradiol	None. Will be in ng/L level if established.
Pharmaceuticals	**Criterion**
Cotinine/Primidone/Phenytoin	1/10/2 μg/L
Meprobamate/Atenolol	200/4 μg/L
Carbamazepine	10 μg/L
Estrone	320 ng/L
Other Chemicals	**Criterion**
Sucralose	150 mg/L
Tris [2-chloroethyl]phosphate	5 μg/L
N,N-diethyl-meta-toluamide	200 μg/L
Triclosan	2,100 μg/L

Source: NWRI 2013

Collecting and analyzing source water quality data is one of the most critical steps when developing brackish water sources and selecting the appropriate concentrate management technology. To provide a reliable basis for planning while controlling the analytical efforts and costs, it is important to understand what to measure, how to measure, at what frequency to measure, and how to analyze the data.

Relevant water quality constituents of concern should include a fully balanced set of cations and anions, including calcium, magnesium, sodium, potassium, strontium, barium, ammonia, alkalinity, sulfate, chloride, fluoride, nitrate, orthophosphate, silica, iron, manganese, and boron, as well as TDS, conductivity, pH, and temperature. Additional water quality data are required to assess the organic and biological fouling potential as well as removal of organics and pathogens, including total or dissolved organic carbon (TOC or DOC), total hydrocarbons, and microbial constituents (total and fecal coliform, heterotrophic plate count, etc.). To assess the propensity of particulate fouling, SDI and turbidity are often used. In wastewater applications, emerging contaminants such as endocrine disruptors and carcinogens (e.g., NMDA) will also have to be measured if the water is intended to be used for indirect potable reuse or DPR. Furthermore, for facilities in operation, chemical usage and operational parameters are also critical for assessing the baseline scenario and characterizing the concentrate quantity and quality.

Assessment of the source water quality is essential for developing a future desalination facility. The characterization of the prospective concentrate stream can be estimated using membrane performance projection software, which is also used to project membrane performance. Supplemented by antiscalant manufacturers' projection models, these tools can be used to quantify the scaling and fouling potential and characterize the concentrate quality. If an existing desalination facility is in operation, water quality monitoring should cover the water quality for not only the feedwater but also the final product and the concentrate. To capture changes in the water quality that will affect planning, design, or operation, historical water quality data over a relatively long period of time should be analyzed instead of relying on limited data sets or a few grab samples. Do not simply assume water quality remains relatively constant without sufficient data to support such an assumption. Ideally, monthly or quarterly data for a full set of pertinent parameters should be analyzed for at least a year. Additionally, for operating facilities, efforts should be made to gather and analyze data from continuous monitoring of feed, product, concentrate conductivity, pH, flow, temperature, etc.

RO AND EDR PERFORMANCE MODELING

Several RO/NF membrane and EDR manufacturers have created and made available membrane system performance projection computer models. These models are available either through download from the manufacturers' websites or through contact with the manufacturers or their representatives (DeMichele at al. 2014). A sample list of available computer models is provided in Table 2-6. The computer models are periodically updated. The latest version of any modeling software should be obtained before developing planning and design criteria for a new desalination process.

Each of these models has a graphical user interface that allows the user to visualize the physical configuration of the system. They are instrumental in

- assessing the feed, product, and concentrate water quality;
- identifying limiting scalant and foulant;
- determining the anticipated recovery of the desalination process; and
- estimating energy consumption for the desalination process.

Supplemented by antiscalant manufacturers' projection models, these tools can be used to quantify the scaling and fouling potential and characterize the concentrate quality. Water quality inputs required for these models typically include TDS, conductivity, pH, calcium, magnesium, sodium, potassium, strontium, barium, ammonia, alkalinity, sulfate, chloride, fluoride, nitrate, orthophosphate, silica, iron, manganese, boron, and SDI.

TREATMENT LEVELS

Comparing raw water quality to the product water quality goals will determine whether desalination technologies are required. Assuming other means (e.g., water conservation, source salinity control, alternative sources, alternative treatment using lime softening or ion exchange) alone cannot meet the product water quality goals, desalination is determined to be inevitable. Concentrate management components must be included to properly handle the residuals generated from the desalination process.

Inland desalination and concentrate management typically involve multiple treatment and disposal steps to treat the various types of source water to meet the product

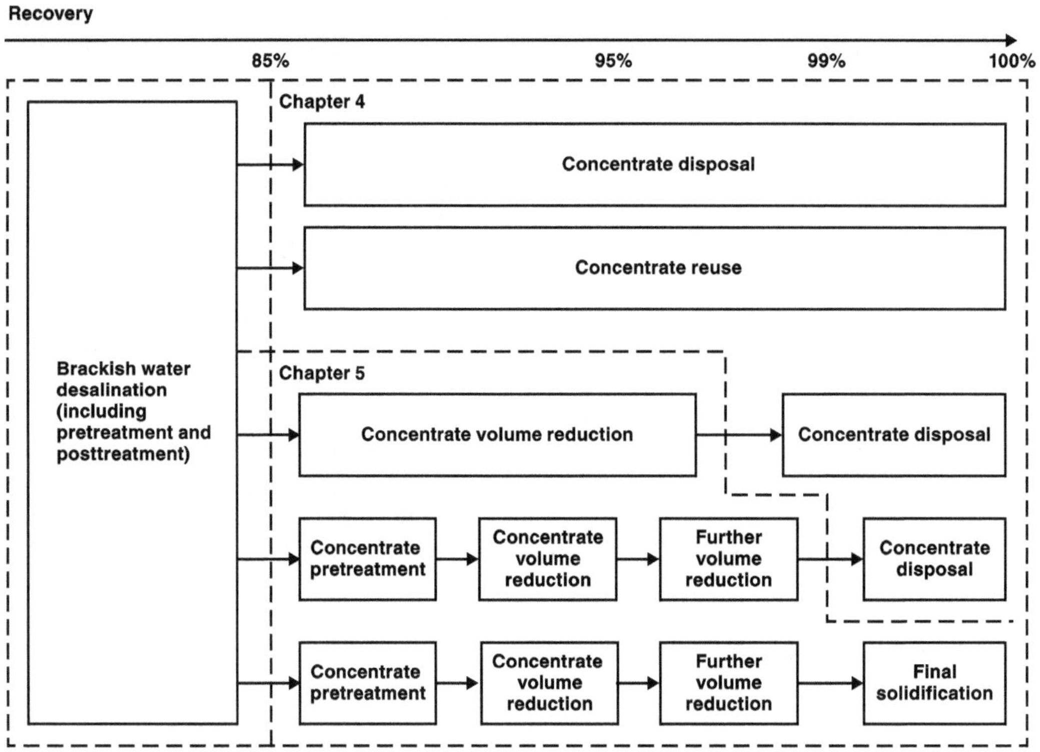

Figure 2-1 Treatment and disposal levels for inland desalination and concentrate management

Table 2-6 Reverse osmosis and nanofiltration design computer models

	Manufacturer	Model Name	Website
RO Projection	CSM	CSMPro V.5.0	www.csmfilter.com
	Dow Water & Process Solutions	ROSA V.8.0.3	www.dowwaterandprocess.com
	Hydranautics	IMSDesign V.2012.8	www.membranes.com
	Koch Membrane Systems	KMSROPRO V.8.06	www.kochmembrane.com
	LGChem (NanoH_2O)	Q+ V.2.0	www.lgwatersolutions.com
	Toray	TorayDS2 V2.0.1.58	www.toraywater.com
	Trisep	TROI	www.trisep.com
	Suez Water Technologies	Winflows 3.1.2	www.suezwatertechnologies.com/resources/winflows
EDR Projection	Suez Water Technologies	Watsys 3.0.30	www.suezwatertechnologies.com

water quality goals. Figure 2-1 illustrates the levels of treatment using a general block diagram. Definitions of these treatment and disposal levels are as follows:

- Brackish water: Water having a mineral content in the range between freshwater and seawater. In water-desalting practice, brackish water is generally considered to be water containing 1,000–15,000 mg/L of TDS.
- Desalination or desalting: A water treatment process, such as distillation, RO, or ED, that removes dissolved mineral salts and other dissolved solids from water.
- Pretreatment: One or more actions or processes occurring immediately before water enters the downstream desalination treatment unit process.

- Post-treatment: The additional treatment of product water (i.e., distillate, permeate, or filtrate) from a desalination system to produce finished water that meets quality requirements, such as permeate post-treatment for corrosion control or disinfection following membrane desalting.
- Concentrate disposal: The process of ultimately discharging a reject stream, most often associated with desalination or membrane processes, to the environment. Concentrate disposal often involves ocean discharge, deep-well injection, or discharge to another receiving medium with a mineral concentration at least as high as the concentrate's concentration.
- Concentrate reuse: The reuse of concentrate for one or more purposes (e.g., dust control, cooling water makeup, wetland or habitat creation), generally only as part of a planned reuse scheme.
- Concentrate pretreatment: One or more actions or processes occurring immediately before concentrate from a desalination treatment process enters the downstream concentrate treatment process.
- Concentrate volume reduction: A concentrate treatment process that aims at reducing the volume of the concentrate by maximizing recovery of product water.
- Further volume reduction: A concentrate treatment process that aims at further reducing the volume of the concentrate by maximizing recovery of product water. This term is often used when multiple concentrate volume reduction technologies are involved in series to achieve high recovery goals.
- Zero liquid discharge (ZLD): Term used to describe the operation of a facility that results in no liquid discharge to a body of water. In desalting, this commonly refers to all liquid streams.
- Final solidification: A concentrate treatment process, such as crystallization, achieving ZLD by converting all dissolved solids in the concentrate into dry solids or crystal form.

TYPES OF BRACKISH WATER DESALINATION TECHNOLOGIES

Understanding the tools in the desalination and concentrate management toolbox helps planners to leverage the strength of each tool when developing site-specific solutions.

Processes used for brackish water desalination can be categorized as follows:

- pressure-driven membranes,
- electric potential driven membranes,
- thermal evaporation,
- ion exchange, and
- other emerging technologies.

Pressure-driven membrane separations include technologies such as RO and NF. These processes use hydraulic pressure to produce purified water from saline feedwater through a semipermeable membrane. Both NF and RO effectively remove inorganic and organic compounds and produce product water of superior quality. NF typically has lower operating pressure requirements than RO. NF is effective in removing TDS, hardness, and constituents with large-size molecules such as arsenic and bromide. But the overall salt rejection and the removal of nitrate, iron, manganese, volatile organic compounds, and emerging contaminates using NF are less than the rejection experienced with RO technologies. These technologies are further discussed in depth in Chapter 3, mainly as primary tools for inland desalination. Coupled with pretreatment addressing the limiting factors, they can also be used for secondary desalination (e.g., concentrate volume reduction, as discussed in Chapter 5). Innovative pressure-driven membrane technologies also include vibratory shear-enhanced processing (VSEP), tubular RO membrane separation, and slurry precipitation and recirculation RO (SPARRO). Various improvements to conventional NF/RO membranes, configurations, and control strategies are also available, such as the closed circuit desalination (CCD) technology or reversible RO.

Electrically driven membrane separation technologies include ED and EDR. This category can be extended to include technologies such as electrodeionization and even nonmembrane processes such as capacitive deionization (CDI). In ED and EDR processes, an electric current migrates dissolved ions through an ED stack consisting of alternating layers of cationic and anionic membranes. Because it does not separate nonpolar substances, it is not as constrained by silica concentration in the feedwater. Because of the open-channel design of ED and EDR membranes, a significant level of suspended materials can be tolerated in the feedwater. However, this process is not a barrier to organics and is less suitable for primary treatment of water containing organics, pathogens, and other nonionic contaminants. A newly developed ED technology, electrodialysis metathesis (EDM), uses four solution compartments and four membranes—rather than two of each in the repeating units for ED or EDR. The employed membranes include both ordinary anion and cation exchange membranes and monovalent selective cation and anion exchange membranes. Chapter 3 discusses these technologies as primary desalination options. Chapter 5 further explores the benefits of using ED and EDR as viable concentrate treatment options for achieving near ZLD concentrate management.

Thermal evaporation–based processes include membrane distillation (MD), multi-effect distillation (MED), vapor compression distillation (VCD), and multistage flash (MSF) distillation. These technologies generally have higher capital and operating and maintenance costs than other desalination processes such as RO. If used for primary desalination, these technologies are generally used for industrial processes, especially when waste heat is available, and could be cost prohibitive for brackish groundwater applications at the municipal scale. However, they remain viable and often crucial for achieving ZLD concentrate treatment.

Emerging technologies include cutting-edge innovations that are currently in the research and development phase, such as membrane desalination cells, aquaporin membrane separation, adsorption desalination (AD), etc. These technologies are not a main focus of this manual but are briefly discussed in Chapter 9, *Technological Trends, Salt Recovery, and Beneficial Uses*.

Table 2-7 presents an estimated range of recoveries, treatment costs, and energy consumptions for treating a given source water using these categories of desalination technologies.

TYPES OF CONCENTRATE MANAGEMENT TECHNOLOGIES

Typical concentrate management technologies are presented in Chapters 4 and 5 of this manual. In Chapter 4, concentrate disposal options and reuse opportunities are discussed, including evaporation ponds, deep-well injection, sewer discharge, and surface water discharge.

In Chapter 5, two main components of concentrate treatment are discussed. First, the concentrate from a brackish desalination process discussed in Chapter 3 may be conditioned, stabilized, or pretreated to address the limiting fouling and scaling issues and to be reduced in volume through secondary desalination technologies such as RO or EDR. This can typically enhance the overall brackish water desalination system recovery from 75–90 percent to 93–95 percent. The second component of concentrate management, referred to as "further volume reduction" in Figure 2-2, aims to increase

Table 2-7 Types of desalination technologies

	Pressure-driven Membrane Separations	Electrically Driven Membrane Separations	Thermal Evaporation Processes	Emerging Technologies
Examples	NF/RO, CCD, SPARRO, VSEP	ED/EDR, EDM, CDI, EDI	MSF, MD, MED, VCD, crystallizer	AD
Major Benefits	• Relatively high energy efficiency • Superior product water quality	• Less susceptible to fouling • Require lesser degree of pretreatment • More flexible operational control of product water TDS	• Ability to use waste heat • Excellent product water quality • Energy consumption relatively insensitive to TDS	• Potential higher recovery and lower energy consumption
Major Limitations	• Energy consumption increases with feed TDS • High scaling and fouling propensity • Require high level of pretreatment	• Energy consumption increases with feed TDS • Poor removal of nonionic contaminants	• Relatively higher energy consumption	• Still in research and development stage

AD—adsorption desalination, CCD—closed circuit desalination, CDI—capacitive deionization, ED—electrodialysis, EDI—electrodeionization, EDM—electrodialysis metathesis, EDR—electrodialysis reversal, MD—membrane distillation, MED—multi-effect distillation, MSF—multistage flash, NF—nanofiltration, RO—reverse osmosis, SPARRO—slurry precipitation and recirculation reverse osmosis, TDS—total dissolved solids, VCD—vapor compression distillation, VSEP—vibratory shear-enhanced processing

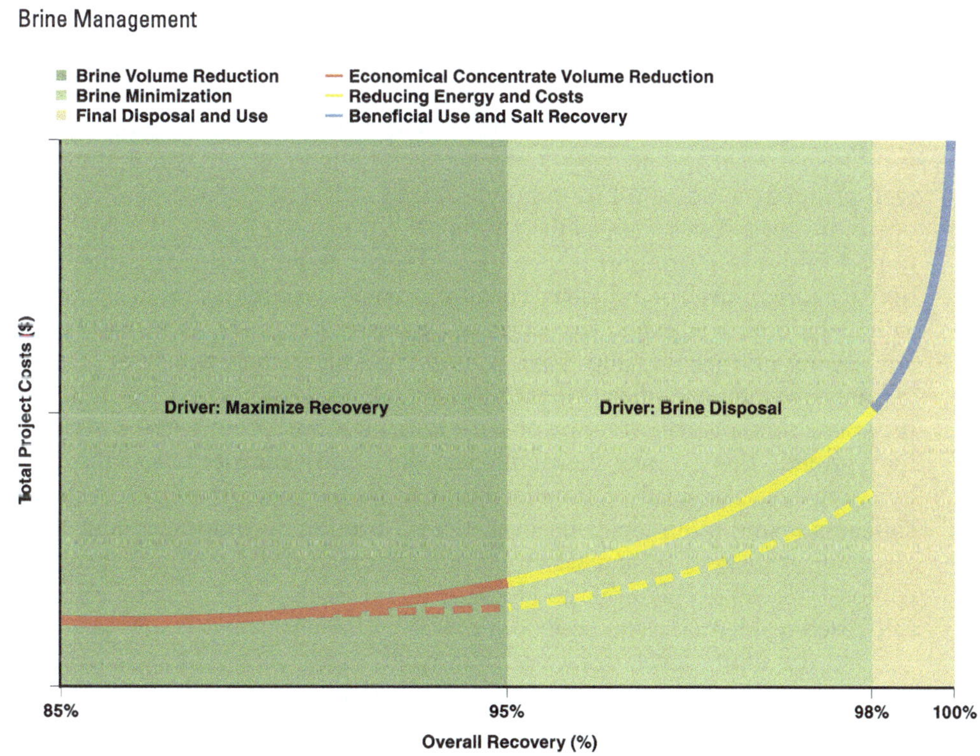

Figure 2-2 Three strategic focuses for inland desalination and concentrate management

overall freshwater recovery to 97–99 percent by using thermal brine concentrators or other innovative technologies such as osmotic brine concentrators incorporating forward osmosis technology. The most advanced brine volume reduction technologies allow achievement of concentrations approaching saturation of sodium chloride at around 200,000–300,000 mg/L TDS. Such brine concentration is typically followed by a final disposal step using small evaporation ponds or a final thermal solidification process using crystallizers and mechanical dewatering of the high-salinity waste stream.

PLANNING STRATEGIES

As mentioned previously, concentrate management is complex, affected by many factors, and involves multiple steps. There is no single technology or solution that can address challenges for all inland communities. Generally, the optimal solution is a combination of several technologies and strategies, customized in a way that site-specific restrictions and resources are considered.

Several key planning strategies for concentrate disposal, treatment, and management are briefly discussed subsequently to provide guidance for utilities to develop customized solutions. These strategies include but are not limited to the following:

- managing salinity proactively, leveraging regional resources, and forming partnerships;
- strengthening water quality databases and establishing an integrated water and salt balance model;
- combining technologies to form customized solutions;
- keeping track of new technologies; and
- tailoring product water quality.

Proactive Salinity Management

As easy as it is for salt to dissolve into water, it is difficult to remove salt from water. Most desalination technologies separate salt from water, which does not eliminate salt and results in a concentrate waste stream. For many inland communities, the concentrate is difficult to treat and manage. Water reuse, desalination, and concentrate disposal, if not well planned, could result in salinity buildup in the recycled stream and contamination of freshwater sources such as the aquifer. If not managed properly, salinity buildup will impact future water resources, resulting in costly treatment before water can be used for human, agricultural, or industrial consumption. Water utilities should proactively manage salinity before the situation worsens and expensive treatment remedies become inevitable.

Many utility companies have water conservation programs that educate customers not only on water use efficiencies but also on salinity issues (e.g., City of Fresno 2007). Such programs promote conservation of salt—using less and putting less down the drain—and may provide tips on how the public can personally contribute to salinity management by using liquid instead of powder laundry soap, using dryer sheets instead of liquid softeners, using mopping pads instead of a traditional mop and bucket of water, and sweeping instead of hosing down an outside area. Many programs promote the use of compost instead of chemicals and fertilizers and recommend more efficient alternatives to salt-based residential water softener units.

At the utility level, proactive salinity management could start with establishing a salinity management plan as part of the system-wide integrated water resource management plan. Early development and adoption of a salinity management plan will allow inland communities to minimize adverse impacts of salinity on economic growth. This effort will help raise awareness of the significance of salinity management, bolster the utilities' long-term efforts, and identify critical gaps and "low-hanging fruit" opportunities. This plan will delineate a roadmap for utilities to manage water portfolios faced with salinity impacts, supporting development and protection of alternative water sources to reduce salinity impacts. Contents of the salinity management plan could vary among communities but typically include adding TDS in the industrial pretreatment program; establishing local limits on sensitivity ions such as sodium, chloride, boron, and selenium; continuing funding; embarking on data collection; planning and protecting new water sources; and considering alternatives to desalination such as water conservation, nonpotable reuse, and indirect and direct potable reuses. Prioritizing industrial development and agricultural activities will help prevent severe environmental issues.

Developing solutions to salinity problems can be challenging and expensive. Utilities should leverage national, regional, and local resources, and consider forming partnerships to combat salinity. Consider joining collaboration efforts such as those led by the Multistate Salinity Coalition, the Texas Water Development Board, the Central Arizona Salinity Study, the Central Valley Salinity Coalition, the Southern California Salinity Coalition, and others. Leverage funding opportunities and federal grants to support salinity management research. Stay connected and form partnerships to develop regional solutions on a larger scale.

Integrated Water Balance Model

Establishing an integrated water and salt mass balance model could provide many benefits for the planning and design of desalination and concentrate management facilities. Unlike the RO and EDR projection models, which are a more detailed analysis of one single process, such a model is designed to help utilities evaluate the system-wide impact and cost implications of salinity. Many planners typically utilize Excel to perform such mass balance calculations. For complex systems, more advanced water system models could offer better platforms to help utilities capture significant savings (Carollo Engineers 2017). This type of model integrates brine treatability evaluation, concentrate management optimization, water resources management, engineering cost estimates, financial analysis, etc., for better decision support. In the big data age, more and more utilities find themselves with a large volume of historical data (e.g., flows, TDS, conductivity, hardness, chlorides, etc.) but do not know how to leverage such information to optimize planning and concentrate management. These advanced decision-support tools fill in the gap. They leverage advanced computing techniques such as Monte Carlo simulation, scenario management (USBR 2012), full factorial analysis, and multi-objective optimization to help utilities develop customized desalination and concentrate treatment solutions that capture million-dollar cost savings (He 2018).

Whether a water and salt balance is established using an Excel spreadsheet or more advanced planning and design software, the ability to manage flow and water quality data—and compare that with demands and product water quality goals—is the first and most critical step toward identifying the concentrate treatment and management requirements, assessing alternative technologies, and developing and optimizing customized solutions. By tracking various flow streams in the whole water cycle as well as TDS and individual ions through the system, the planners and engineers can evaluate multiple desalination and concentrate alternatives side by side, optimize the configurations of the desalination and concentrate management systems, and tailor the product water quality to meet the needs in the most efficient way.

Combining Technologies to Form Customized Solutions

Because the source water qualities and the tolerance of the local environment to salinity and individual ions are different for each inland community, no single technology or any fixed combination of technologies can fit all brackish water desalination needs. As presented in Figure 2-1, many levels of treatment technologies and strategies must be involved to form a site-specific solution. Understanding the mechanism, strength, and limitations of each technology helps to leverage and combine them to achieve an overall solution. While the later chapters of this manual will provide more in-depth discussion on individual technologies, this section presents several strategies and principles on combining them into customized solutions.

As illustrated in Figure 2-2, the total project costs and efforts associated with inland desalination and concentrate management increase exponentially as the overall recovery increases. On this curve, three critical areas of concentrate management—brine volume reduction, brine minimization, and final disposal and use—should be given strategic emphasis when combining technologies to develop customized brine management solutions.

First, with the improvements in desalination technologies and the pretreatment, enhanced recovery (typically 85–95 percent) could be achieved relatively economically. Concentrate from the primary desalination steps could be reduced by half or up to two-thirds when applying technologies such as lime softening (Gabelich 2007), ion exchange softening, or organic pretreatment (Carollo Engineers 2013) before the secondary desalination using RO, NF, ED, or EDR.

One example to illustrate the value of this strategy is given subsequently. The recovery of brackish water RO is often limited by sparingly soluble constituents, such as silica, barium, strontium, and calcium. To maximize RO recovery, the limiting scale-forming constituents must be identified, removed, and/or controlled. This can be accomplished by maximizing scaling control using an optimized scale inhibitor dosing strategy (i.e., high doses of innovative scale inhibitor products with optimized pH control). However, the effectiveness of scale inhibitors could still be limited. For example, the RO recovery for a brackish groundwater desalination application in Phoenix, Ariz., was limited to 85 percent. When operating at 90 percent recovery with a dendrimer-type scale inhibitor at the maximum dose, the same RO plant performed better initially but showed scaling/fouling in the RO membranes after less than 30 days. Periodic membrane cleaning was not able to sustain permeate production (Carollo Engineers 2006).

Alternatively, removing the scale-forming constituents through a chemical precipitation (i.e., lime softening) step allowed higher recovery through a subsequent secondary RO system. For the same Phoenix application discussed previously, when calcium carbonate, magnesium hydroxide, and silica are removed using a solids-contact clarifier, an overall system recovery of 94–95.5 percent was achieved using a subsequent secondary RO system to recover more water from the softened concentrate (Carollo Engineers 2009, He 2010). Additional information about such technologies is provided in Chapter 5.

Second, achieving higher recovery (>95 percent) would require additional efforts to address several limiting factors. As the cost for producing water using these technologies are too high compared to conventional water production or desalination technologies, the main goal becomes brine minimization in lieu of enhancing recovery. The focus is to find technologies that can reduce the brine volume more economically than ZLD technologies such as thermal brine concentrators and crystallizers.

To achieve this goal, the power of multiple processes may have to be combined to address the limitations of a single technology. Using this principle, many commercialized concentrate treatment packages are basically combining various pretreatment and desalination processes. For example, the High Efficiency Reverse Osmosis (HERO™) process uses lime softening and ion exchange to pretreat the feed for primary and secondary RO to maximize recovery. GE's AquaSel™ system combines EDR and chemical precipitation to recover approximately 95 percent or more product water. In this system, the EDR concentrate recirculation stream is diverted to a chemical precipitation reactor, where the limiting scale-forming ions (e.g., barium, calcium) are removed through chemical softening using chemicals such as lime, caustic soda, and/or soda ash. The softened concentrate is then returned to the EDR feed. By addressing the limiting constituents in the chemical precipitation step, higher recovery can be achieved without scaling the membranes.

Third, costs required to further reduce the concentrate volume and achieve ZLD become even higher. To offset such costs, attention is often given to turning the waste into a resource and recovering salt products from the last one or two percent of the concentrate. Technically, this is achievable but the feasibility highly depends on the marketability of the recovered products. Relevant technologies are discussed in Chapter 9 of this manual.

Keeping Track of New Technologies

Continuing innovations in desalination and concentrate management technologies provide opportunities for improved performance with reduced cost and energy consumption. In the past several decades, better antiscalant and antifoulant agents, more efficient pretreatment systems, and better membranes have been lowering the barrier for implementation of concentrate management facilities. Newly developed nanomodified membranes improved the energy efficiency of RO membranes while reducing the fouling potential. Mechanical improvements were implemented for conventional systems to improve the ability to handle higher concentrations and more aggressive applications. More advanced control strategies and system optimization, such as RO and CCD, provide opportunities to reduce desalination costs and concentrate volume for certain applications. New operation strategies such as predictive cleaning methodology reduce overall operation costs (Chesters et al. 2013). Refer to Chapter 9 for more details on such trends and developments.

Desalination and concentrate management often involve complex and expensive treatment and require long-term planning. Inland utilities combating salinity challenges should act proactively, act early, and keep track of the latest developments in desalination and concentrate management technologies. Although not all of these innovations are applicable for all communities, keeping apprised of industry trends and current best practices helps in formulating the right customized solution to local salinity challenges. Utilities should participate in local or regional salinity research and accumulate firsthand experience and data to prepare them for solving the complex challenge of salinity management. Given that the planning phase for a desalination plant often extends beyond five to ten years, technologies that are currently in the research and development stage could be mature and ready to implement as the planning and implementation phase proceeds.

Tailored Water Quality Using Technologies with Better Selectivity

Waters of differing quality are needed for different end uses, including drinking and cooking, irrigation of yards and public parks, urban lake replenishment, toilet flushing, groundwater recharge to replenish local groundwater, and water recycling for cooling towers and various industrial uses. Desalination technologies such as RO are effective in removing TDS and most contaminants but often cannot remove the target contaminants selectively. By tailoring water quality to meet different needs, utilities may capture cost-saving opportunities to reduce the use of RO and the generation of concentrate, conserve energy, facilitate local water reuse, promote nutrient recovery, and have an overall lower carbon footprint.

When groundwater contains inorganic contaminants but the TDS is not high, RO or other desalination technologies may not be the best solution. This could also apply to developing non-RO–based treatment trains capable of producing water suitable for DPR. When applicable, utilities should consider employing alternative water treatment technologies (e.g., ion exchange, adsorption, EDR, softening, coagulation, granular media filtration, biological filtration, granular activated carbon, advanced oxidation) that can be used to replace, minimize, or supplement concentrate-generating technologies such as RO.

For reclaimed water recycling, tailored water quality makes sense when certain nonpotable reuse demands can be met without treating the product water to drinking water quality—at least not 100 percent of the stream. For example, if the product water is for irrigation, removing all ions and nutrients using RO and then applying more chemicals and fertilizers could be more sustainable. Also, water with lowered sodium concentrations is more suitable for irrigation, and water with lowered hardness is more ideal for cooling towers. To accomplish this, selected treatment technologies must be flexible enough to selectively remove target ions. In the planning and development of desalination and concentrate management facilities, technologies with the ability to selectively remove target contaminants deserve priority consideration, as they could help utilities meet the ultimate water supply goals more economically. Finding ways to tailor water quality to meet all needs requires sound engineering judgment and tools such as the advanced integrated water and salt balance model mentioned previously.

CONCLUSIONS

The following summarizes the key points discussed in this chapter:

- Evaluate the brackish source water quality and quantity as well as the characteristics of the contaminants limiting concentrate management options.
- Establish the most suitable treatment goals for the site-specific conditions of a given project.
- By comparing the water quality with the treatment goals, identify treatment requirements and limiting factors and provide a basis for the development of planning strategies.
- Several key planning strategies for concentrate disposal, treatment, and management can help utilities to develop customized solutions to salinity issues, including establishing a salinity management plan; collecting relevant water quality data; developing integrated water and salt balance models; and combining multiple technologies to overcome concentrate management challenges.
- Utilities should leverage national, regional, and local resources and form partnerships to combat salinity.

REFERENCES

Antony, A.; Low, J.H.; Gray, S.; Childress, A.E.; Le-Clech, P.; and Leslie, G. 2011. Scale Formation and Control in High Pressure Membrane Water Treatment Systems: A Review. *Journal of Membrane Science*, 383(1-2):1–16.

AWWA (American Water Works Association). 2007 (2nd ed.). M46, *Reverse Osmosis and Nanofiltration*. Denver: AWWA.

AWWA. 1995. M38, *Electrodialysis and Electrodialysis Reversal*. Denver: AWWA.

Ayers, R.S., and Westcot, D.W. 1985. Water Quality for Agriculture. *FAO Irrigation and Drainage Paper*, 29:1

Barclays Official California Code of Regulations, Title 22. Social Security, Division 4. Environmental Health, Chapter 15. Domestic Water Quality and Monitoring Regulations, Article 16. Secondary Drinking Water Standards. §64449 Secondary Maximum Contaminant Levels and Compliance. https://govt.westlaw.com/calregs/Document/I2260318DFFF045529B9496276F3A8573?transitionType=Default&contextData=(sc.Default) (accessed Feb. 25, 2019).

Carollo Engineers. 2017. *Blue Plan-it® Decision Support System Fact Sheet*. www.carollo.com/blueplan-it (accessed Nov. 30, 2018).

Carollo Engineers. 2009. *City of Phoenix Western Canal WTP Water Quality Sampling and Testing Project Final Report*. Walnut Creek, Calif.: Carollo Engineers.

Carollo Engineers. 2006. *City of Phoenix USBR Phoenix Area Membrane Pilot Study Phase I and Phase II Reports*. Walnut Creek, Calif.: Carollo Engineers.

Chesters, S.P.; Armstrong, M.W.; Fazel, M.; Wilson, R.; and Golding, D.A. 2013. *RO Membrane Cleaning, Past, Present, Future—Innovations for Improving RO Plant Operating Efficiency*. Proc. IDA World Congress on Desalination and Water Reuse, Tianjin, China.

Chesters, S.P.; del Vigo, F.; and Darton, E.G. 2007. Theoretical and Practical Experience of Calcium Phosphate Inhibition in RO Waters. Proc. IDA World Congress, Maspalomas, Gran Canaria, Spain.

City of Fresno (Calif.). 2007. *The Seriousness of Salt*. http://www.cvsalinity.org/index.php/docs/documents-and-presentations/65-city-of-fresno-salinity-brochure/file.html (accessed Apr. 25, 2018).

Clark, D.R.; Green, C.J.; and Gordon, J.A. 2000. Laboratory Exercises to Demonstrate Effects of Salts on Plants and Soils. *Journal of Natural Resources and Life Sciences Education*, 29:41–45.

DeMichele, D.; Seacord, T.F.; and Sutherland, RO. 2014. *Manual of Practice for the Use of Computer Models for the Design of Reverse Osmosis/Nanofiltration Membrane Processes*. Austin, Tex.: Texas Water Development Board.

Gabelich, C.J.; Williams, M.D.; Rahardianto, A.; Franklin, J.C.; and Cohen, Y. 2007. High-Recovery Reverse Osmosis Desalination Using Intermediate Chemical Demineralization. *Journal of Membrane Science*, 301(1):131–141.

Gallagher, C.D. and Dietrich, A.M. 2010. TDS and Temperature Affect Consumer Taste Preferences. *Opflow*, 36(12):20–22.

He, C. 2018. You Can Bet on It: Optimizing Desalination Facility Sizing Using Monte Carlo Simulation. Proc. 2018 Membrane Technology Conf., West Palm Beach, Fla.

He, C. and Kelso, B. 2010. Pilot Studies Examine RO Concentrate Management Strategies. *WaterWorld*, 26(12).

Maas, E.V. 1984. Salt Tolerance of Plants. In *Handbook of Plant Science in Agriculture*. Boca Raton, Fla.: CRC Press.

Ning, R.Y., A. Tarquin, M. Trzcinski, and G. Patwardhan. 2006. Recovery Optimization of RO Concentrate From Desert Wells. *Desalination*, 201(1–3):315-322.

NWRI (National Water Research Institute). 2013. Examining the Criteria for Direct Potable Reuse. Fountain Valley, Calif.: NWRI.

Sheikholeslami, R. and Tan, S. 1999. Effects of Water Quality on Silica Fouling of Desalination Plants. *Desalination*, 126(1–3):267–280.

USBR (US Bureau of Reclamation). 2012. *Colorado River Basin Water Supply and Demand Study.* https://www.usbr.gov/watersmart//bsp/docs/finalreport/ColoradoRiver/CRBS_Executive_Summary_FINAL.pdf (accessed Oct. 30, 2018).

USBR. 2003a. *Central Arizona Salinity Study, Phase I, Technical Appendix J: Economic Assessment Model.* https://www.usbr.gov/lc/phoenix/programs/cass/pdf/Phase1/JTechapdxEconomics.pdf (accessed June 29, 2018)

USBR. 2003b. *Central Arizona Salinity Study, Phase I, Technical Appendix L: Reported Impacts of High Salinity Water on Golf Courses in Central Arizona.* https://www.usbr.gov/lc/phoenix/programs/cass/pdf/Phase1/LTechapdxGolfcourses.pdf (accessed Apr. 25, 2018).

USEPA (US Environmental Protection Agency). 2017. *Potable Reuse Compendium.* www.epa.gov/ground-water-and-drinking-water/2017-potable-reuse-compendium (accessed Feb. 18, 2019).

WateReuse Research Foundation (WRRF). 2015. Framework for Direct Potable Reuse. Project 14–20. Alexandria, Va.: WRRF.

WHO (World Health Organization). 2017. *Potable Reuse: Guidance for Producing Safe Drinking-Water.* http://apps.who.int/iris/bitstream/handle/10665/258715/9789241512770-eng.pdf;jsessionid=328ECBC3213E0B0577FC1A37EBBA5D98?sequence=1 (accessed Apr. 25, 2018).

Chapter 3

Brackish Water Desalination

Val S. Frenkel and Greg Wetterau

This chapter presents an overview of brackish water desalination, brackish water characteristics, and commonly used technologies for the treatment of brackish water supplies. The section is focused on the primary desalination technologies (e.g., RO and EDR), including pretreatment and post-treatment polishing or stabilization. It does not cover additional technologies involved with concentrate reduction or elimination of liquid waste streams, which are discussed in Chapters 4 and 5.

There are numerous methods that can be used for the desalination of brackish water. While the selection of the most appropriate technology should be made based on environmental, water quality, and cost considerations for each specific application, membrane technologies are the most commonly used for brackish water treatment. The two most commonly used methods of brackish water desalination include pressure-driven desalination technologies, such as RO and NF, and electric potential–driven desalination technologies, such as ED and EDR. Each will be discussed further in this chapter.

Other technologies employing various methods of distillation, forward osmosis, and chemical-based desalination processes are used less commonly as the primary desalination technology to treat brackish water supplies; these are discussed in later chapters of this manual in relation to concentrate recovery methods.

The choice of desalination technology will influence both the quality of the treated product water (e.g., TDS, microbial content, trace organic compounds) and the treatment processes required upstream (pretreatment) and downstream (post-treatment) of desalination. This chapter will discuss pretreatment and post-treatment in relation to each technology.

PRESSURE-DRIVEN MEMBRANE DESALINATION

Process Description

RO and NF currently represent the state-of-the-art of desalination technology being used for the vast majority of all brackish water desalination applications. This is because their ability to reject a variety of contaminants ensures that most of the treatment objectives can be met in a single process, with lower energy consumption, lower feedwater flows, and no thermal impact in the concentrate discharge (as are common for thermal desalination processes). Improvements in membranes have continued to decrease the energy required to desalinate water while reducing the fouling rate of the membranes. Moreover, RO and NF are established technologies that are commercially offered by several vendors.

RO and NF membranes create two streams from a source water. Water that passes through the membranes is referred to as permeate and is the treated water with most dissolved ions removed. A waste stream containing rejected ions and water that has not passed through the membranes is referred to as concentrate or reject. The general characteristics of RO and NF membranes—including the physical structure of the membranes, treatment efficiencies, recovery rates, energy requirements, and membrane fouling—are discussed subsequently.

Structure of Membranes

RO and NF are pressure-driven membrane processes in which the feedwater is pressurized to exceed its own osmotic pressure. The water then passes through a dense membrane, which is engineered to retain salts and low-molecular weight solutes. The first RO plant was constructed in Coalinga, Calif., and began operating in 1965 using commercial RO membranes like those developed at UCLA (Loeb and Johnson 1967). The original asymmetric cellulose acetate (CA) membranes, developed in the 1960s, were less permeable than modern thin-film composite (TFC) polyamide membranes and required higher driving pressure. Additionally, the ability of the original CA membranes to reject salts was significantly less.

Figure 3-1 illustrates the structure of the TFC membrane where a dense, thin polyamide membrane layer and thicker, porous polysulfone and polyester support layers are separately engineered to provide necessary support against high feed pressures without creating excessive headloss to water passing through the membrane surface.

Today, there are also a variety of modified and improved blends of CA membranes available to the desalting industry, but these membranes are rarely used in large-scale

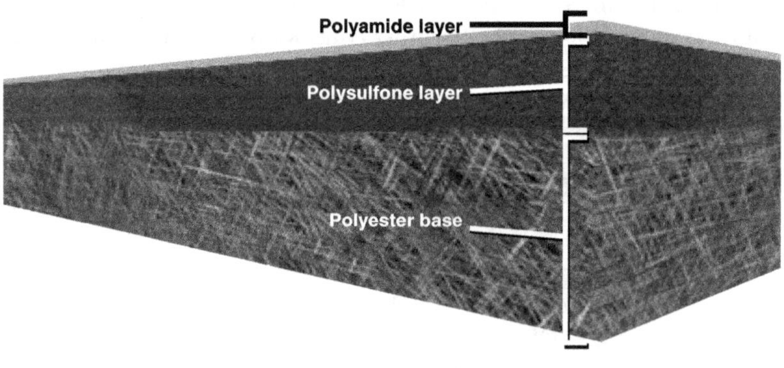

Adapted with permission from Hydranautics

Figure 3-1 Composite polyamide membrane structure

desalination applications. These CA membranes can tolerate continuous exposure to low concentrations of chlorine (0.1–0.5 mg/L at 25°C), which is an advantage for biofouling control in brackish surface water applications. However, they are susceptible to hydrolysis, which compromises the membrane's salt rejection performance. Hydrolysis of CA membranes is accelerated if the operating pH is greater than 6.5, requiring acidification for most brackish water applications (Mallevialle et al. 1996).

The development of TFC membranes provided greater salt rejection and water production. The TFC membranes are made by forming a thin, dense, solute-rejecting surface film on top of a porous structure. The material of construction for the top surface film layer is mostly aromatic polyamide (PA) and the bottom support layers are polysulfone and polyester (see Figure 3-1). The thin, nonporous top layer contributes to contaminant rejection, and the porous bottom layers provide the mechanical strength. Because of the thin top layer, pressure required to drive water through the membrane material is significantly reduced. For example, Water Factory 21, which was operated in Orange County, Calif., using CA membranes for treating high-salinity wastewater, had a design pressure of 500 psi. Today, RO facilities using modern TFC membranes at the same location have a design pressure that is less than half this value.

The TFC membranes are stable over a broad pH range (2–11) and can withstand temperatures as high as 45°C. However, unlike the CA membranes, they are susceptible to degradation by free chlorine. Although the rate at which membranes succumb to attack by free chlorine is a function of pH, their performance gradually (and sometimes catastrophically) deteriorates upon exposure. That being said, there are several facilities that operate TFC membranes on brackish feedwater that contains 2–5 mg/L of chloramine residual to control biofouling on the membrane surface.

Osmotic Pressure and Reverse Osmosis

Using membranes to desalinate high-salinity water requires forces strong enough to overcome natural phenomena such as osmotic pressure. Natural osmotic pressure will push water from the low-salinity side of a semipermeable membrane to the high-salinity side to balance the ionic strengths of the separated solutions (as shown in Figure 3-2 from right to left). RO is the process by which external pressure is applied to the high solute concentration, driving water back to the low solute concentration (as shown in Figure 3-2 from left to right), while retaining ions within the high solute concentration. The leakage of the ions from high to low solute concentration is caused primarily by the diffusion. The higher the salinity gradient across the membrane, the higher the osmotic forces that must be overcome.

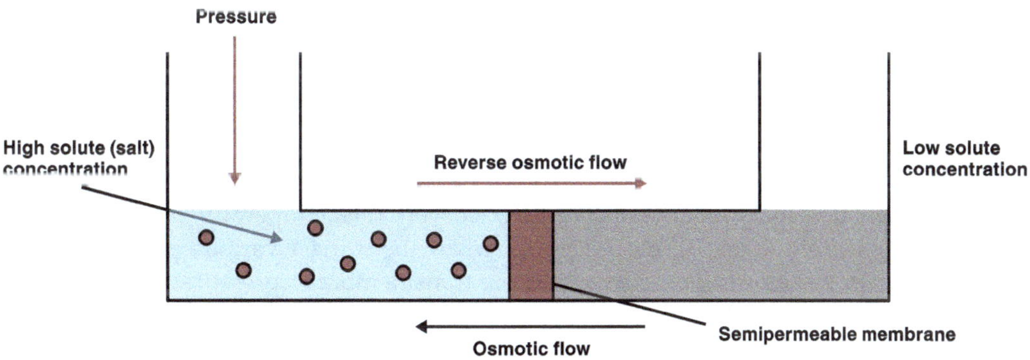

Figure 3-2 Basic concept of natural osmosis and reverse osmosis

The osmotic pressure can be approximated for brackish water supplies using a modified version of the Van 't Hoff equation, as presented in Equation 3-1.

$$P_{osm} = 1.12 \times (T + 273) \times \sum m_j \qquad \text{(Eq 3-1)}$$

where

P_{osm} = osmotic pressure (psi)
T = temperature (C)
m_j = sum of molarity concentration of all constituents (moles/L)

The numerical coefficient used in this equation (1.12) incorporates various conversion factors, thermodynamic constants, and empirical factors but will differ depending on the specific constituents present in the water. The version presented in Eq 3-1 is used by Dow Corporation in their RO design literature, while a coefficient of 1.19 is used by Hydranautics. Actual equations used by the RO vendors in their projection software are more complex and differ from vendor to vendor (Wolfe et al. 2011).

Because osmotic pressure is directly proportional to the ionic strength of a water being treated, a higher-salinity brackish water will require higher pressure and higher energy to overcome the osmotic forces. For example: A brackish water with salinity of 1,000 mg/L could result in a source water osmotic pressure of 9 psi, which would increase to 27 psi in the RO concentrate if operated at a recovery of 70 percent, requiring a feed pressure of approximately 140 psi. In contrast, a brackish water with salinity of 8,000 mg/L could have an osmotic pressure in the source water of 83 psi, which would increase to 270 psi in the concentrate if operated at the same conditions, requiring a feed pressure of 322 psi. For this reason, it is generally less costly to treat lower-salinity water than it is to treat higher-salinity water.

Treatment Efficiency

The USEPA has designated RO as a best available technology for removal of numerous inorganic contaminants, including antimony, arsenic, barium, chloride, fluoride, nitrate, nitrite, boron, selenium, radionuclides, TDS, and constituents of emerging concern, including endocrine-disrupting compounds (synthetic and natural hormones) and several pharmaceutical compounds.

Current brackish RO membranes remove between 98 and 99.2 percent of the TDS from the source water, while NF membranes have a lower removal rate, with removal efficiency heavily affected by the charge and molecular weight of the constituents present in the water. NF membranes have higher rejection capabilities for divalent ions such as calcium, magnesium, and sulfate. A few analytes, such as boron and nitrate, are only moderately rejected by RO membranes, while many dissolved gases, such as hydrogen sulfide, carbon dioxide, and lower molecular weight volatile organic compounds, have little or no reduction, even by the tightest RO membranes. RO and NF membranes will also remove viruses, bacteria, and protozoa, with greater than 3-log reduction demonstrated in multiple studies (Pype et al. 2013, Mi et al. 2004).

Removal efficiency of dissolved salts will be affected by numerous operating parameters, including membrane flux, recovery rate, temperature, pH, membrane fouling, and membrane age. Table 3-1 summarizes the impact various parameter changes will have on the removal efficiency and the transmembrane pressure required to push water through the membranes. Membrane flux is defined as the permeate production rate per unit membrane area; in the United States, it is usually expressed as gal/day-ft^2 (gfd). The most widely adopted metric units are LMH (liter/m^2/hour).

Table 3-1 Impact of operational changes on removal efficiency and pressure

Parameter Change	Salt Removal Efficiency	Transmembrane Pressure
Increasing flux	Increases	Increases pressure
Increasing recovery	Decreases	Increases pressure
Increasing temperature	Decreases	Typically decreases pressure, but can increase pressure when osmotic pressure is high
Membrane fouling	Typically decreases removal, but can temporarily increase removal with some types of biological fouling	Increases pressure
Membrane age	Decreases	Increases pressure
Oxidant damage	Decreases	Decreases pressure

Nanofiltration Versus Reverse Osmosis

RO and NF membranes are similar in many ways and even frequently referred to as one technology—"NF/RO"—especially in early planning stages. Both use pressure-driven semipermeable membranes to remove dissolved solids. Both can be fouled or scaled and could benefit from pretreatment. However, they can be distinguished based on the size and charge of contaminants that each is able to remove. For example, RO membranes are very efficient at removing most ions, with the exception of some low-molecular weight, weakly charged molecules and dissolved gases (boron, hydrogen sulfide, carbon dioxide). Typically, NF membranes are capable of removing larger and more strongly charged ions but allow a higher percentage of smaller molecular weight and monovalent ions such as sodium, nitrate, and chloride to pass through.

A relatively recent technology, NF was developed primarily as a membrane-softening process to offer alternatives to chemical softening. It is also often used to control disinfection byproducts precursors or remove color. The commercially available TFC NF membranes are characterized by a porous (around 2 nm pore diameter) and negatively charged skin layer. Because of the negative charge on the membrane material, divalent hardness ions are removed to a greater extent than are monovalent ions (e.g., 95 percent calcium removal and 70 percent sodium removal). The higher concentration of single charged ions on the permeate side also helps reduce the osmotic backpressure. This, combined with a loose membrane skin layer, reduces the hydraulic pressure requirements considerably; however, the salt rejection for these membranes is often insufficient for treating many brackish water sources.

Design Considerations

Membrane configurations. RO and NF membranes are most commonly used in a spiral-wound configuration, where multiple flat sheet membranes are wrapped together into a single membrane module or element. Figure 3-3 presents a schematic representation of a membrane module. Each membrane sheet is glued at the edges to an identical sheet, creating a membrane envelope, with purified permeate collected on the inside of the envelope and the feed and concentrate on the outside. A thin mesh permeate spacer is placed inside the envelope to allow free flow of the product water into a permeate collector tube at the center of the module. Feedwater flows tangentially along the surface of the membrane, through feed spacers used to separate membrane sheets. As the feedwater travels along the length of the membrane, pressure drives the RO process, forcing water through the dense membrane layer.

The feed becomes more concentrated as it passes along the surface of the membrane, creating a concentrate stream that exits the membrane module on the opposite side from

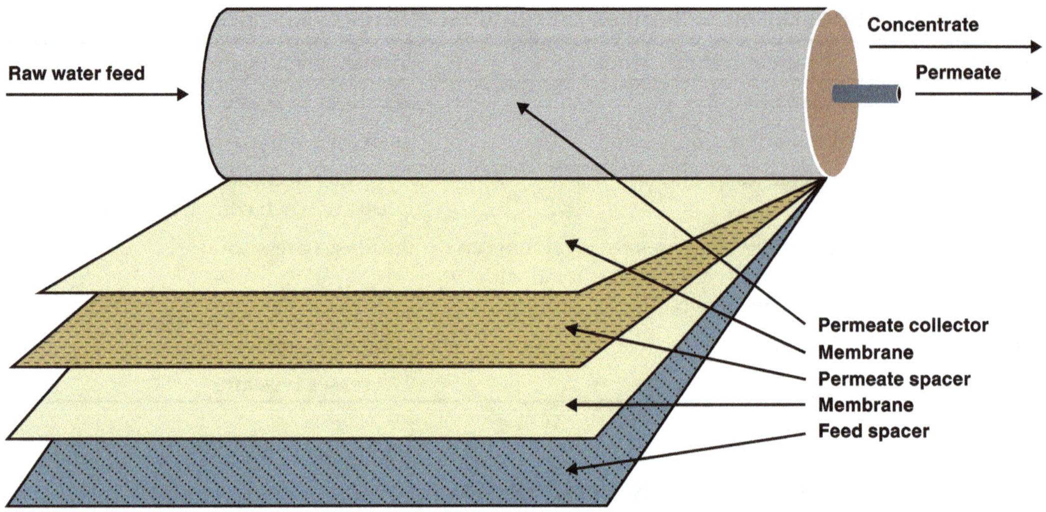

Figure 3-3 Schematic representation of a spiral-wound membrane module

the feed. This crossflow configuration helps to remove concentrated salts from the surface of the membranes, reducing concentration polarization effects, the rate of fouling, and salt passage into the permeate.

Multiple membrane modules are typically placed in series within a fiberglass pressure vessel. Typically, six to eight modules are used in a single pressure vessel. Multiple pressure vessels are then configured in parallel into a larger membrane skid. Figure 3-4 shows RO membranes being loaded into a pressure vessel on a membrane skid. Couplings are used to connect the permeate tubes of adjacent membrane modules, with an end cap adapter connecting the first and last module to permeate ports at the ends of the pressure vessels. The integrity of the o-rings in these connection points is essential for maintaining separation between the purified permeate flow and the concentrate stream. Various integrity-monitoring methods are used to ensure that all membrane seals are intact within an operating system.

Currently, the standard RO membrane diameters on the market are 2.5, 4, and 8 in. The large-diameter RO membranes were introduced at 16 and 18 in. The length of RO modules was standardized to 40 and 60 in. (61 in. for large-diameter RO).

In brackish RO facilities, it is common to use two and sometimes three stages to increase the water recovery as much as possible. In these configurations, two first-stage pressure vessels will typically feed a single second stage, with the concentrate from the first stage becoming the feed for the second. Reducing the number of pressure vessels in the downstream stages allows the crossflow velocities within the membranes to remain high enough to provide adequate removal of salts from the membrane surfaces and maintain minimum velocities recommended by the membrane manufacturer. Figure 3-5 shows a schematic representation of a two-stage RO process.

Two-stage systems should not be confused with two-pass configurations, used commonly in seawater desalination. In a two-pass configuration, the permeate from the first pass is sent through a second pass for additional treatment to improve the product water quality for specific constituents not sufficiently removed in the first pass. In contrast, a two-stage configuration is used to increase overall recovery, producing more product water from the concentrate stream of the first stage. Third and fourth stages can also be used to further reduce the concentrate flow stream; however, a point will eventually be reached at which either the concentration of sparingly soluble salts or the osmotic pressure will prevent further reduction of concentrate flow.

Figure 3-4 Eight-inch RO membrane modules being loaded into a membrane pressure vessel

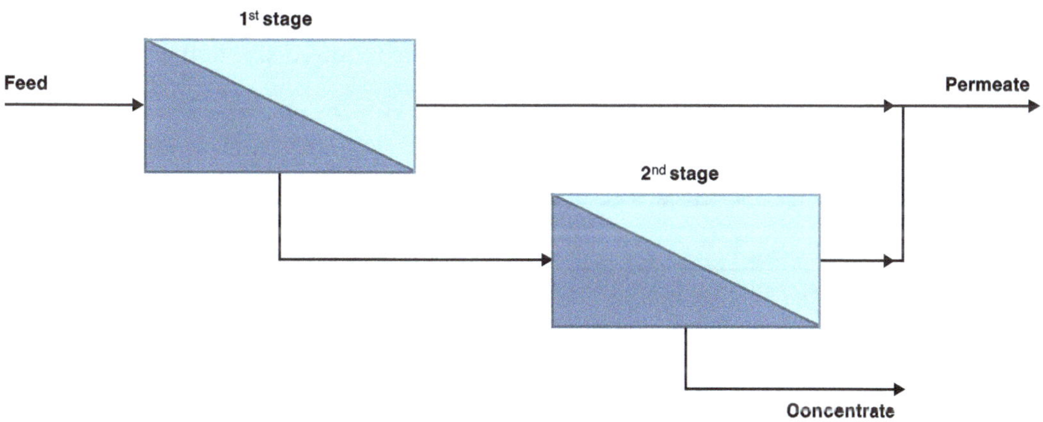

Figure 3-5 Schematic representation of a two-stage RO process

Recovery rates. Recovery as a hydraulic parameter is defined as the total amount of water produced from the total amount of water used. This can be calculated as the ratio of the permeate flow to the feed flow. Recoveries in brackish water RO applications typically range from 50 to 85 percent, with higher recoveries achieved with three- and four-stage RO designs, which are discussed later in this manual. The maximum recovery will be dictated by the concentration of sparingly soluble salts, methods used to control scaling, and the overall salinity of the water.

Balancing flux. Pressure is lost in the feedwater as it passes along the membrane surfaces from the feed to the concentrate of each element and from one element to another, leaving a lower driving pressure in the tail element compared to the driving pressure in the first. While the residual pressure decreases, the concentration of salts increases, creating a larger osmotic pressure that must be overcome to continue producing permeate. Therefore, permeate flux through the tail element will be lower than permeate flux through the first element. Available pressure will still be lower in the feed to the second stage, with the increasing osmotic pressure decreasing production further. The lower flux and higher concentration of salts will also decrease the quality of product water from the second stage, i.e., result in higher permeate TDS.

Flux balancing can be used to increase production from the second-stage elements while decreasing production from the first. The benefits of flux balancing include a higher overall flux for the entire system, a decreased risk of fouling on the first stage, and improved overall water quality in the blended permeate. The primary drawback is a higher capital or operating cost, depending on the method used for flux balancing. Flux balancing can be achieved by providing an interstage pressure boost, throttling the first-stage permeate, or selecting looser membranes for the second stage that will naturally balance flux between stages. Table 3-2 presents methods of flux balancing and the advantages and disadvantages of each.

Energy recovery. As stated previously, RO facilities require significant energy because of the high driving pressures required to overcome osmotic forces in the brackish water. Much of this energy remains in the concentrate stream leaving the last stage of membranes. This energy is often dissipated over a concentrate control valve or fixed orifice plates. Energy recovery devices have been developed that make use of this residual energy to boost the pressure of either the primary feed or the interstage feed, reducing the overall energy required for the desalination process. Energy recovery devices have become commonplace in seawater desalination, with devices recovering more than 98 percent of the residual energy and drastically reducing the required size of the primary feed pumps.

Table 3-2 Methods used for balancing flux between stages

Flux Balance Approach	Advantages	Disadvantages
Permeate throttling	Simplest method of flux balancing No additional equipment required Often used for pilot plants and small systems	Least efficient method Increased operating cost for head burned across valve Risk of damage from reverse flow through membranes during shutdowns
Interstage booster pump	Efficient method of flux balance Flexible operation over range of salinities and pressures Does not require permeate throttling	Additional equipment and piping adds to capital cost Less efficient than energy recovery devices
Energy recovery device	Uses residual pressure from concentrate to boost interstage pressure Most efficient method of flux balancing	Higher capital cost than interstage booster pump Devices optimized for small range of flows and pressures Limited interstage boost achieved for low-salinity waters
Lower rejection membranes in second stage	No additional equipment required for flux balancing Natural flux balance does not require operator attention to maintain	Decreased product water quality with lowest-rejection membranes treating poorest-quality water

However, in brackish water desalination, less energy is present in the concentrate streams, and the energy recovery devices tailored for these applications are considerably less efficient than the devices used in seawater applications. While energy recovery has been used in brackish water designs for more than 25 years, energy recovery devices have not yet become standard in those designs because of the relatively low energy recovery rate and cost compared to the cost of additional equipment.

Energy recovery devices can be classified as either centrifugal or positive displacement. Centrifugal devices include the basic reverse running pump (or Francis turbine), the higher-efficiency Pelton wheel turbine, and the direct coupled turbocharger. Positive displacement devices include pressure exchangers and work exchangers, which tend to be high in efficiency but are often less effective in low-pressure conditions or when large disparities in flow exist between the high-energy waste stream and the low-energy feed. These devices are therefore used primarily in seawater applications, where waste flows and waste energy are highest.

To date, the direct coupled turbocharger is the primary type of energy recovery device used in brackish water applications. This device uses the concentrate flow to drive a turbine directly coupled to a pump on either the feed or interstage flow. These devices are most frequently used to boost the interstage flow, providing a method of flux balancing that does not require external energy inputs. This configuration is shown schematically in Figure 3-6. Figure 3-7 is a photo of a turbocharger energy recovery device in use at a brackish water desalination facility.

Turbocharger devices typically have efficiencies near 50 percent, considerably lower than the efficiencies of positive displacement devices used in seawater desalination. The devices must be optimized for a narrow range of flows and pressures, which can result in greater inefficiencies when pressures change because of variations in salinity, membrane fouling, or aging of membranes. Turbocharger devices have therefore been developed with external motors that can provide additional energy input, allowing the device to be operated over a larger range of conditions with a more consistent pressure boost to the second stage.

Positive displacement energy recovery devices that are targeted for use in brackish water applications have recently been developed. While their use remains somewhat

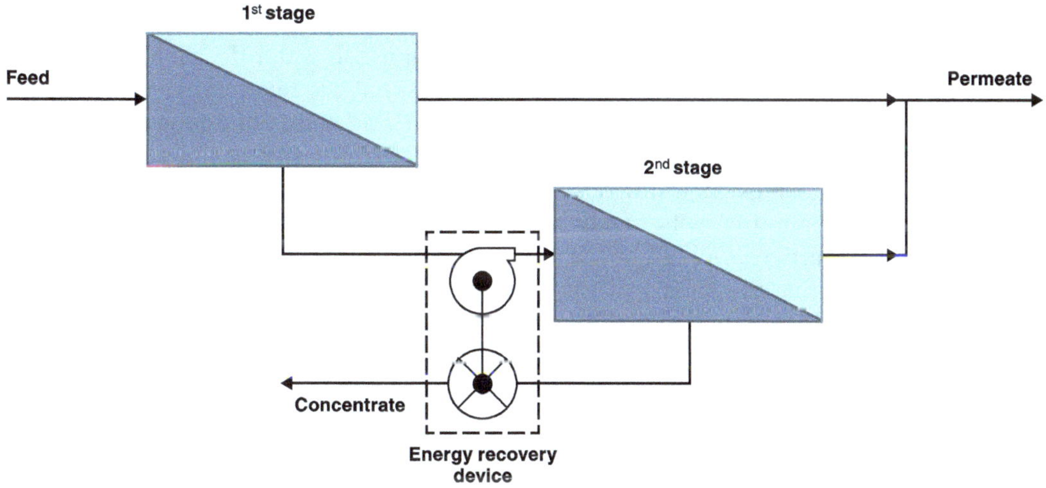

Figure 3-6 Schematic representation of a two-stage RO process with a turbocharger energy recovery device

Figure 3-7 Energy recovery device installed in a brackish water desalination facility

limited, they may provide a more efficient alternative for energy recovery at future brackish water desalination facilities.

Integrity monitoring. The majority of brackish water desalination plants currently in operation treat well supplies from aquifers that are relatively deep and disconnected from surface water supplies, where pathogenic microbial pollutants are a significant risk. Because of the low risk of pathogens in these groundwater supplies, integrity testing methods at brackish water desalination facilities have historically not been driven by regulatory requirements but by practical measures for maintaining the product water quality objectives. Online conductivity analyzers have been the most commonly used tools for monitoring the integrity of the desalination process. These analyzers serve both as a method of identifying impacts of membrane fouling and as a warning of breached o-rings, seals, or compromised membranes that allow excess passage of salts into the permeate stream. Most RO and NF facilities also allow sampling from individual pressure vessels as well as probing of the permeate tubes to help identify and locate specific integrity failures.

As the industry has grown, desalination facilities have been constructed on an increasing number of brackish surface water supplies and on shallow wells that are more directly influenced by surface waters and more susceptible to microbial pathogens. With development of these newer brackish water supplies, surface water treatment regulations have begun to be applied to an increasing number of desalination facilities, prompting the development of more sensitive integrity monitoring approaches to ensure consistent removal of pathogens. Early studies evaluating the effectiveness of RO systems to remove pathogens found a wide variation in removal effectiveness for MS2 virus, ranging from 99 percent (2-log) to 99.9999 percent (6-log). These early studies did not see the microbial removal effectiveness correlating well with rejection of surrogate compounds, such as sodium chloride (Adham et al. 1998, Gagliardo et al. 1997). More recently, however, with improved methods of quality control in manufacturing, and with an increased focus on maintaining system integrity in the design and operation of desalination facilities, subsequent studies have demonstrated that greater than 3-log reduction of pathogens can be consistently maintained in an RO system (Pype et al. 2013, Mi et al. 2004).

While high levels of pathogen removals can be demonstrated in controlled studies, online integrity monitoring methods remain somewhat limited for brackish water desalination facilities. The low-pressure microfiltration/ultrafiltration hollow fiber membrane filtration facilities typically use offline pressure decay testing to demonstrate greater than 4-log pathogen reduction and identify integrity breaches in the membranes. Because of the structure of the composite membranes and the configuration of the spiral-wound elements, these methods cannot be employed with existing brackish water membranes. Online monitoring methods that can maximize the pathogen reduction credits granted to an RO system are therefore being sought.

In California, one brackish water desalination facility has been granted 2-log pathogen reduction credits using online conductivity monitoring as a surrogate (Wetterau et al. 2011); however, this facility uses seawater desalination membranes, which achieve a higher rejection of TDS than brackish water desalination membranes can achieve. Similarly, potable reuse facilities have been granted up to 2-log pathogen reduction credits using online TOC monitoring as a surrogate (Wetterau et al. 2013). This alternative could be used at brackish water desalination facilities with relatively high TOC levels in their source water, but 99 percent reduction will be more challenging to demonstrate for supplies with lower TOC because of limitations in the currently available online instruments. One brackish water desalination facility in Colorado has been granted 2.5-log pathogen reduction credits using intermittent grab samples for sulfate as a surrogate (Rynders et al. 2013), and a potable reuse facility in Australia has been granted 3-log pathogen credits using a similar approach (Moscovis 2013).

Numerous studies focused on identifying reliable integrity monitoring approaches for RO are currently underway. As the potable reuse industry continues to grow and as brackish water RO continues to be employed on an increasing number of surface water–influenced supplies, it is anticipated that integrity monitoring approaches will become more standardized and will be better correlated with the actual pathogen-removal effectiveness of the membranes.

Chemical cleaning. RO and NF systems are equipped with clean-in-place (CIP) systems to allow periodic chemical cleaning of the membranes and piping with acid, base, or proprietary cleaning solutions. A chemical feed pump and storage tank form the main components of the CIP system, and a tank heater is often used to improve cleaning effectiveness. The CIP system will typically contain the following components:

- booster pump,
- CIP tank with heater,
- chemical feed pumps to add cleaning chemical to the tank,
- flowmeter, and
- cartridge filter.

The CIP process is intended to remove scale and organic and biological foulants from the membranes and hydraulic piping. The chemical solutions circulated through the membranes depend on the type of contamination. This is a required procedure used in preventive maintenance that should be completed according to the manufacturer's recommendations.

Only the chemical solutions approved by the membrane manufacturer should be used in the CIP process to avoid irreversible damage to the membrane. Temperature and pH limits, as provided by the membrane supplier, must be strictly followed. The following chemical solutions are commonly used.

Acid solution. Periodic cleaning with a two percent citric acid solution (pH 3–4) is the most commonly used method for removal of inorganic, nonsulfate scale as well as metal oxides (iron and manganese). While other acids, such as phosphoric, sulfuric, or hydrochloric, are also used, the organic content in citric acid has been found to improve cleaning effectiveness, serving as a surfactant as well as an acid. Mixtures of citric acid and stronger acids are sometimes used when a lower pH (1–2) is required to remove inorganic scale; however, cleanings should never be done at a pH lower than the recommended limit from the membrane manufacturer. Acid solutions are sometimes heated, although common scales such as calcium carbonate and magnesium hydroxide can be removed more effectively at lower temperatures.

Alkaline solution. Caustic soda is used for pH cleaning (pH 10–12) to remove organic and microbial foulants as well as sulfate scale. Alkaline cleanings are most commonly done at elevated temperatures (35–40°C) to improve the effectiveness of foulant removal. When surfactants or biocides are used to enhance RO cleaning, they are most commonly used in conjunction with the alkaline cleans. Different membranes have different limits on pH and temperature. It is therefore critical that cleaning procedures be tailored to the specific membranes in use at each facility.

Proprietary cleaning solutions. Proprietary cleaning solutions are often used in cases of excessive biofouling or for removal of silica or other difficult-to-remove foulants. These solutions employ either detergents and surfactants to enhance removal of organic material or biocides to reduce or control biological growth. Proprietary cleaning solutions are often combined with alkaline cleaning agents and are most commonly used at elevated temperatures (35–40°C).

Safety Considerations

Proper operation of an RO or NF system requires an understanding of the safety issues involved in high-pressure pumping, chemical handling, and mechanical systems. Each of these is discussed briefly here.

High-pressure pumping. While pressures used in brackish RO are considerably lower than those used in seawater desalination, feedwater pressures still exceed 100 psi and can exceed 300 psi in higher-salinity applications. System equipment components should therefore be designed to have a relatively long useful life (15–30 years) to the extent that is possible and practical. All plant components should be designed such that failures are repairable. Suppliers should provide documentation on individual equipment components that verifies that the components are suitable for the water quality stated in the specifications.

The structures and equipment components of a desalination system should be constructed of materials that are corrosion resistant in the environment in which they are placed. High-quality stainless steel and plastics may be acceptable materials of construction provided they can withstand the rigors of their internal and external environments.

Many products and components may be used in systems in which pressures are other than ambient. In such instances, the materials of construction must meet the codes established by federal, state, and local regulatory agencies and professional associations such as the American Society of Mechanical Engineers.

Chemical safety. Hazardous chemicals, including acids and sodium hydroxide, are commonly used at brackish water desalination facilities, requiring proper precautions to prevent the risk of exposure to plant personnel. Personnel should always wear protective clothing (rubber gloves, apron, face shield, goggles, boots, and chemical-protection suit) when handling corrosive solutions or when working on the chemical pumps and supply lines. In addition, they should always wash down the work area after handling the chemicals and always clean up any chemical spills. When acid and water are being mixed, it is important to always add the acid to the water. Adding water to acid can cause a violent explosion, spraying acid on the operator. When possible, it is recommended to avoid adding sulfuric acid to pressurized feed lines to minimize the potential impact from leaks and the required pressures in the chemical injection system.

Mechanical systems. Centrifugal and vertical turbine pumps should never be operated with either the discharge or suction valve closed. This can cause friction to the impeller and heat the trapped volume of water to a high temperature. If this heated water escapes, it can cause serious burns.

It is also important to never operate any rotating equipment (specifically pumps, blowers, or fans) without their protective guards in place.

REVERSE OSMOSIS PRETREATMENT

The major purpose of RO pretreatment to condition water is to minimize adverse impact on RO operation. Pretreatment is required to minimize the fouling potential of the feedwater to the RO membranes. Pretreatment generally does not have an impact on the product water quality, with the exception of acid addition, which will decrease the pH of the treated water and generally requires post-treatment to address. Pretreatment for brackish water desalination facilities has traditionally been limited because of the low fouling potential of most groundwater sources. The most common methods of pretreatment include chemical addition to prevent scaling on the membranes and cartridge filters for removing particulates that can plug membranes. In some cases, more extensive pretreatment may be needed to reduce sand loading from wells, control iron and manganese, or address high fouling potential with a surface water source. The following includes a brief discussion on RO pretreatment selection and design criteria, including traditional pretreatment, sand control, iron and manganese control, and integrated membrane systems for fouling control.

Traditional pretreatment. As stated previously, chemical addition and cartridge filters are the most commonly used methods of pretreatment for brackish water RO. Acid and antiscalant are injected before the cartridge filters to condition the water and prevent scaling on the membranes.

Antiscalants, also known as scale inhibitors or threshold inhibitors, are organic polymers that inhibit precipitation of sparingly soluble salts, such as silica, carbonates, and sulfates. Typical doses of antiscalant range from 2 to 6 mg/L. The compounds are generally nonhazardous but may be moderately acidic. By using antiscalants, RO facilities are able to operate at higher recoveries without the concentrate exceeding the normal solubility limits of these sparingly soluble salts. Proprietary antiscalant compounds are available from multiple vendors and are often targeted for prevention of specific scale types. Some vendors have developed software to support selection of the most appropriate antiscalant for a given application; however, most facilities still rely on pilot testing or full-scale demonstration testing to select an antiscalant compound. Most antiscalant vendors will perform membrane autopsies to identify the types of scale present on the membranes and recommend an antiscalant product to address this scale.

Acid addition is used to supplement the antiscalant and prevent the formation of pH-dependent scale, such as calcium carbonate and calcium phosphate. Acid is injected most commonly to maintain the Langelier Saturation Index (LSI) in the concentrate below 2–2.5, as higher LSI values have been found to overwhelm the antiscalant and cause significant scaling on the membranes. LSI is commonly used in the industry to characterize certain components' oversaturation by RO in brackish water. Sulfuric acid is the most commonly used acid in municipal RO facilities, but hydrochloric acid has also been employed in locations where sulfate scale is a significant concern. Both acids are extremely hazardous and cause strong exothermic reactions when blended with water. Concentrated acid solutions will react with most metals and plastics, requiring special care in the selection of the storage tanks, piping, and metering pumps. Acid doses vary substantially, depending on the target pH and source water quality characteristics but are generally between 20 and 120 mg/L, with the target pH of the feedwater typically between 6.5 and 7. Because of the danger of handling acid, some brackish water RO facilities have eliminated or avoided using it, relying instead on antiscalant and/or reduction of the recovery rate to prevent scale formation.

Cartridge filters are used at most brackish water RO facilties to remove particulates that will plug or foul the membrane elements. Cartridge filters will remove more than 90 percent of the particulates larger than the pore size of the filters, which is typically between 1 and 10 μ. Cartridge filters rely on depth filtration to trap particulates either on the outside of the cartridge filter or within the filter media itself. The filters cannot be backwashed or chemically cleaned and are replaced when the headloss across the filters exceeds 15–25 psi.

Cartridge filters can be either wound or of melt-blown construction. Wound filter elements are the least costly and most commonly used. They are typically made from polypropylene strings wound around a central core, although nylon, glass fiber, polyester, and other materials are sometimes used. Melt-blown cartridge filters are more costly than wound filters but provide better removal rates for particulates and are less likely to see particulate unloading when differential pressures increase. Similar to wound elements, melt-blown elements used for RO pretreatment are most commonly of polypropylene construction. The US municipal RO industry has generally standardized around 40-in.-long cartridge filter elements with a 1-in. inside diameter. Multiple filter elements are housed within a single filter vessel, and both horizontal and vertical vessels are available. Figure 3-8 shows an installation with multiple horizontal cartridge filters treating an RO supply.

Sand control. Some RO facilities have run into problems with sand production from wells that can overwhelm the cartridge filters and break through to the RO elements. Sand from wells often can be addressed through well rehabilitation; however, some RO facilities have also installed sand separators or strainers to remove sand before it reaches the cartridge filters. When strainers are used, they are typically self-backwashing wedgewire screens with a slot size between 100 and 500 μ. Some RO facilities have also employed vortex-type sand separators.

Iron and manganese control. Iron and manganese present in groundwater supplies commonly contain only reduced forms of the metals (Fe^{2+} or Mn^{2+}), which are highly soluble and do not create a risk of fouling on RO membranes. Iron and manganese can create problems with RO operation if the metals become oxidized, causing fouling from the low-solubility metal oxides. Brackish water RO facilities are operated successfully on waters with iron or manganese exceeding 2 mg/L without additional pretreatment by preventing any oxidation of the metals before their removal by the membranes. At some facilities, however, oxygen cannot be prevented from entering the system, and the iron and/or manganese must be removed before reaching the RO membranes. Removal of iron is far simpler than manganese removal and can be achieved with an oxidation/filtration pretreatment step, using aeration or chlorine as the oxidizing agent and media filters as the

Figure 3-8 Horizontal cartridge filter vessels used as RO pretreatment

filtration step. Iron should be removed to below detection limits, to the extent possible, to prevent membrane fouling.

An oxidation/filtration step can also be employed at locations where high levels of manganese are present and oxidation cannot be prevented, but some type of catalyzing medium is often needed to achieve complete removal of the manganese. Greensand or manganese dioxide can be used for this, but other proprietary media are also available. Employing chlorine oxidation ahead of RO membranes creates a risk of membrane damage if the chlorine is not entirely removed before reaching the membranes.

Integrated membrane systems. Integrated membrane systems using microfiltration (MF) or ultrafiltration (UF) as pretreatment for RO have been employed for treatment of brackish surface waters but are unlikely to be needed for groundwater supplies. MF and UF provide near-complete removal of bacteria and algae, which can cause biofouling on RO membranes. Because of the high fouling potential in the source water, wastewater RO facilities have generally standardized pretreatment around integrated membrane systems, with the use of chloramines also serving to prevent regrowth of microorganisms after their removal through MF or UF. In brackish water desalination, chloramines are not generally used but can be employed as an additional measure for biofouling control.

Integrated membrane systems can also be designed in a direct-coupled configuration, without a break tank between the UF and RO. By limiting the residence time between

Figure 3-9 The Taunton River desalination plant in Brockton, Mass., uses integrated membrane systems to treat a brackish surface water supply.

the UF and RO, the risk of biological regrowth after the UF membranes is significantly reduced. MF and UF membranes may be either submerged or pressurized. An example of an integrated membrane system is shown in Figure 3-9. For more information on MF and UF membrane systems, please refer to AWWA M53, *Microfiltration and Ultrafiltration Membranes for Drinking Water*.

REVERSE OSMOSIS POST-TREATMENT

Post-treatment after RO may be required for removal of dissolved gases (such as hydrogen sulfide), stabilization of the product water, or additional disinfection. RO product water is typically very aggressive, with pH less than 6 and very low alkalinity and hardness. Post-RO adjustment of pH is needed to prevent corrosion of piping, tanks, and pumps. RO permeate from brackish water wells is typically high enough in dissolved carbon dioxide to necessitate air-stripping towers to increase the pH. These towers are typically referred to as degasifiers in Florida and other eastern states and as decarbonators in California and other western states. The stripping towers remove the dissolved carbon dioxide, typically increasing the pH to above 8.0 without chemical addition. Stripping towers are a low-cost means of increasing the pH, but they will not add alkalinity to the water and are typically not able to stabilize the product water without additional post-treatment. For well supplies with elevated levels of hydrogen sulfide, stripping towers have a secondary purpose of removing the sulfides that pass through the RO membranes.

Groundwater desalination facilities commonly use raw water blending to provide a stable hardness in the product water. Blend ratios vary from 5 to 70 percent, depending on the salinity of the source water and the finished water quality goals. At facilities where blending cannot be used, hardness must be increased through the addition of either lime or calcium chloride. Furthermore, caustic soda is commonly injected to reach a pH of 8.5 as a final method for product water stabilization.

Sodium hypochlorite is generally injected into the RO permeate to provide secondary disinfection or a redundant primary disinfection step. When chlorine is used for primary disinfection, a 4-log virus reduction is typically required. This will require a chlorine dose and contact time (CT) of 6 mg/L-min for a temperature of 10°C or 4 mg/L-min at 15°C.

ELECTRODIALYSIS DESALINATION

The ED and EDR systems are designed to remove dissolved minerals from water by the migration of ions through a stack of anion and cation membranes under the force of a direct current (DC) electric field. The ED and EDR processes are similar technologies: while ED process is using DC not flipping polarity, EDR is flipping polarity. Flipping polarity means periodically flipping positive and negative charge of electrodes to minimize scaling of the membranes as ED/EDR is using the alternating current (AC) power supply.

Process Description

Overview. An overview of a water treatment spectrum based on size and molecular weights of contaminants is shown in Figure 3-10 (Frenkel 2007, 2009). ED processes remove TDS in the form of ions or salinity in the range similar to that of RO or NF. RO operates by passing water through the membrane and rejecting dissolved contaminants, whereas ED processes operate by passing ions (charged compounds) through the membranes (as water flows parallel to the membranes). As a result, ED systems are generally more robust with respect to fouling and scaling than RO, and ED systems can generally be designed for higher recovery than RO systems for certain applications.

Figure 3-11 shows the electrodialysis process, the basis for ED and EDR technologies. An electrodialysis stack is composed of two electrodes, a negatively charged cathode and a positively charged anode, as well as repeating pairs of ion exchange membranes (cation and anion).

ED and EDR are membrane-separation processes driven by an electrochemical potential difference induced by an applied voltage that separate dissolved ions from a fluid (water, wastewater, or industrial process fluid). The ED/EDR processes use a voltage potential, instead of pressure, to drive the ions through semipermeable membranes. The positive ions move toward cation-transfer membranes, while negative ions move toward anion-transfer membranes. ED/EDR can be used on brackish water sources and have a high removal efficiency for multivalent ions such as calcium and magnesium (hardness); they will also remove monovalent salts such as sodium and chloride depending on the voltage potential and the selectivity of the ED/EDR membranes.

Membrane Characteristics

The ion selectivity (selection of specific ions for removal) and electrical conductivity are the most important properties of an ion-exchange membrane. Ion selectivity of cation

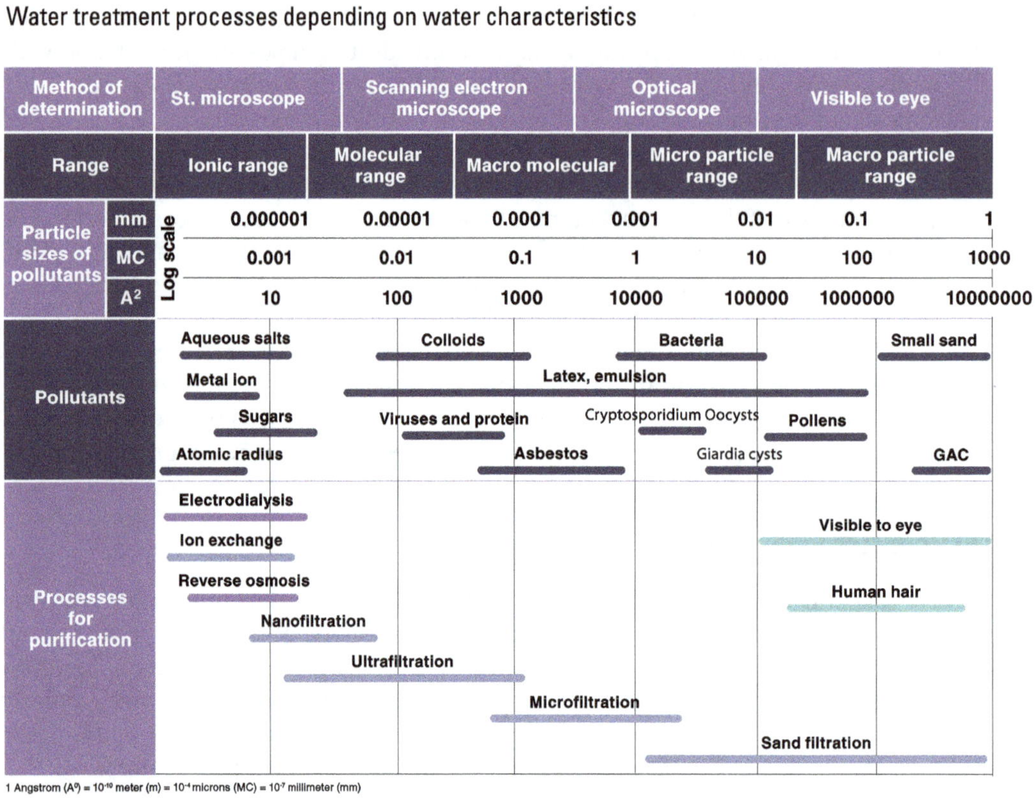

Adapted with permission from Val S. Frenkel

Figure 3-10 Water treatment spectrum

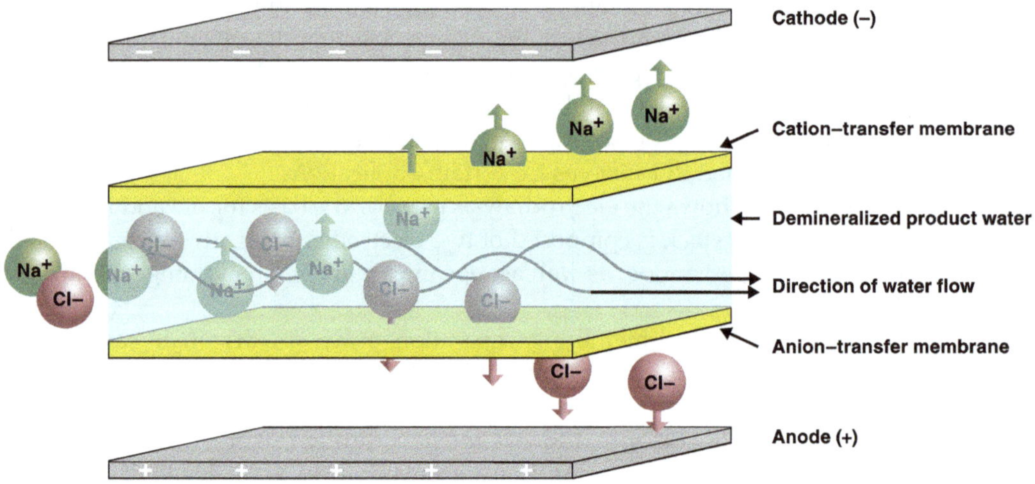

Adapted with permission from Val S. Frenkel

Figure 3-11 Electrodialysis process schematic

and anion membranes is greater than 90 percent for solutions with a normality of less than 0.5*N* up to 0.7*N*, but ion selectivity decreases at higher concentrations. Ion-exchange membranes that have a low electrical resistance consume less energy, which increases system efficiency. As the concentration of the solution increases, resistance decreases. The organic content of source water is an important factor in membrane selection. Organic acids, usually present in brackish well water, decrease membrane efficiency. As they adsorb to the anion exchange membrane, they increase membrane resistance and lower ion transfer. The subsequent need for cleaning can be decreased with the selection of appropriate membranes (Hamada 1993).

Ion-exchange membranes used in ED are described as either anion transfer or cation transfer membranes. These membranes are flat with a smooth surface, similar in appearance to a plastic sheet and to RO membranes. Impermeable to water under pressure, membranes are reinforced with synthetic fiber. Both anion and cation transfer membranes have the following characteristics:

- low electrical resistance insolubility in aqueous solutions,
- semi-rigid construction for ease of handling resistance to change in pH from 2 to 9,
- ability to function continuously at ambient temperatures above 115°F (46°C),
- resistance to osmotic swelling (volume expansion due to water uptake) in solutions between 220 and 60,000 mg/L salt,
- long life expectancy (average 10 years for anion and 15 years for cation membranes),
- resistance to fouling,
- impermeability to water under pressure, and
- availability in many sizes and configurations.

The thickness of ED membranes is from 0.004 to 0.024 in. (from 0.1 to 0.6 mm), and the membranes are either homogeneous or heterogeneous. Membranes are manufactured by mixing an ion-exchanging resin with a polymer. The ion exchanger must constitute at least half the polymer mixture. This mixture is then heated and poured into a supporting open-weave cloth. A copolymerization reaction takes place when homogeneous membranes are produced. Dissolved ions transfer through the membrane in either direction.

Cation membrane. The cation membrane is a sheet in which a cation exchange resin is embedded. This membrane allows the passage of cations (positive) while blocking the passage of water and anions (negative). Negatively charged resin repels the negative ions while allowing the positive ions to pass through the membrane. Figure 3-12 shows the chemical structure of a typical cation membrane. As feedwater travels along the membrane,

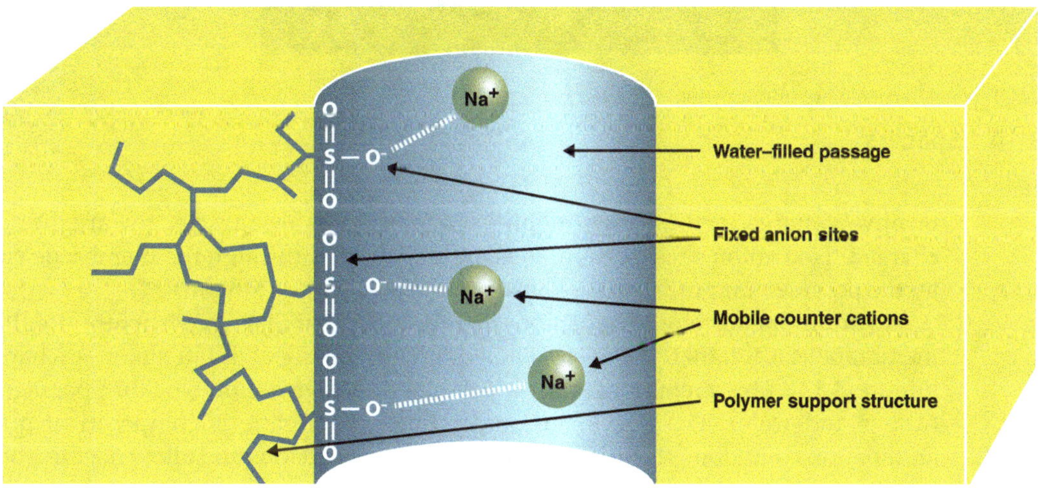

Adapted with permission from GE Water Technologies (formerly Ionics Inc.)

Figure 3-12 Structure of a cation membrane

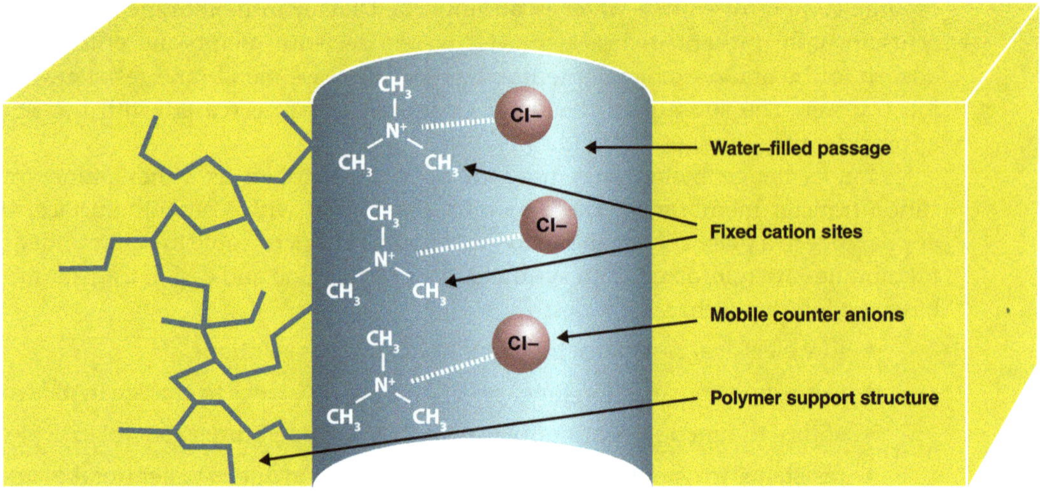

Adapted with permission from GE Water Technologies (formerly Ionics Inc.)

Figure 3-13 Structure of an anion membrane

Courtesy of GE Water Technologies (formerly Ionics Inc.)

Figure 3-14 Spacers

positively charged sodium ions (mobile counter cations) are attracted toward the negatively charged fixed anion sites. Sodium ions are transported through the membrane supported by the polymer support structure and into the concentrate stream.

Anion membrane. The anion membrane is similar in structure to the cation membrane except that the resin embedded in the sheet is an anion exchange resin (Figure 3-13). This membrane allows passage of anions while blocking passage of both cations and water. The active site on an anion membrane is composed of quaternary ammonium compounds, which repel positively charged ions and allow negatively charged ions to pass through the membrane.

Spacers. The ED/EDR spacers are die-cut polypropylene flow channels with manifold cutouts similar to those for membranes (see Figure 3-14). Manifold cutouts form pipes in

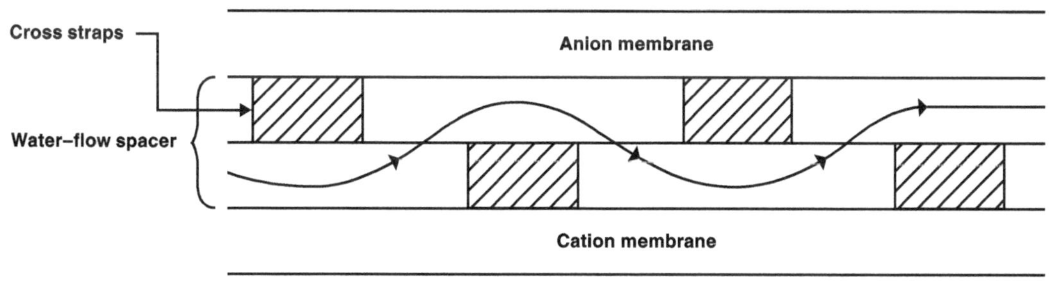

Adapted with permission from Ionics Inc.

Figure 3-15 Baffled flow pattern created by spacer

the stack, allowing water to flow between membranes. Spacers formulated of polyethylene netting perform identical functions. Spacers are alternately positioned between membranes in the stack to create independent flow paths and support membranes.

Demineralizing and concentrating spacers are created by rotating an identical spacer 180°. Demineralizing spacers allow water to flow across membrane surfaces where ions are removed, whereas concentrating spacers prevent the concentrate stream from contaminating the demineralized stream.

A two–plies–thick spacer forces water to follow a baffled pattern around the cross straps, which are pieces of polyethylene placed along the length of the spacer, as shown in Figure 3-15. Increasing turbulence in this manner promotes mixing of the water, efficient use of the entire membrane area, and the transfer of ions. Turbulence resulting from spacers also breaks up particles or slime on the membrane surface and attracts ions to the membrane surface. Flow velocity ranges from 7 to 14 in./s (from 18 to 35 cm/s), creating a pressure drop between the inlet and outlet. A velocity less than 7 in./s (18 cm/s) promotes polarization, or the point of limiting density of water. Maximum pressure for ED and EDR systems is generally limited to less than 50 psi (345 kPa), and pressure is lost at each stage of the system. Because pressure must be maintained throughout the system, the impact of spacers on pressure is an important design consideration. The higher the turbulence, the higher the pressure drop, and vice versa.

Spacers create flow paths within the membrane stack for demineralized and concentrated water. Different models and sizes of spacers satisfy specific design applications. The main difference in spacer models is the number of flow paths, which determines water velocity across the membrane stack and contact time of the source water with the membrane. Water velocity is responsible for the degree of mixing and the amount of desalting that occurs across membranes, making velocity an important design parameter for spacer choice. Because the same spacers are used for both demineralized and concentrated water in EDR systems, the flow rates of both these streams should be equalized to prevent high differential pressures across the membranes (Siwak 1993).

Spacer design and flow velocity are also important parameters in determining the maximum amount of desalting or the current density (discussed later) at which a given system can operate.

Design Considerations

ED/EDR systems are designed to provide the required level of demineralization by staging the membrane stacks in series. Staging provides sufficient membrane area and retention time within the stack to achieve the desired level of demineralization. There are two types of staging in ED/EDR: hydraulic and electrical.

Hydraulic staging. Hydraulic staging refers to increasing the number of passes the water makes along a membrane stack in an ED or EDR system. Maximum salt removal for any hydraulic stage is 50–67 percent, depending on the manufacturer, with typical design values of 40–65 percent. Increased salt removal is accomplished by the addition of more hydraulic stages.

In systems that require higher capacities, additional stages are created by the addition of more membrane stacks (as shown in Figure 3-16). In small systems, hydraulic staging can be accomplished in a single stack by using one or more thick interstage membranes, as shown in Figure 3-17.

Electrical staging. Electrical staging refers to the number of electrodes in a membrane stack. Normally, there is only one electrode pair in a membrane stack, but a second independent electrode pair can be inserted to provide maximum salt removal while avoiding polarization and hydraulic pressure limitations (which are explained later in

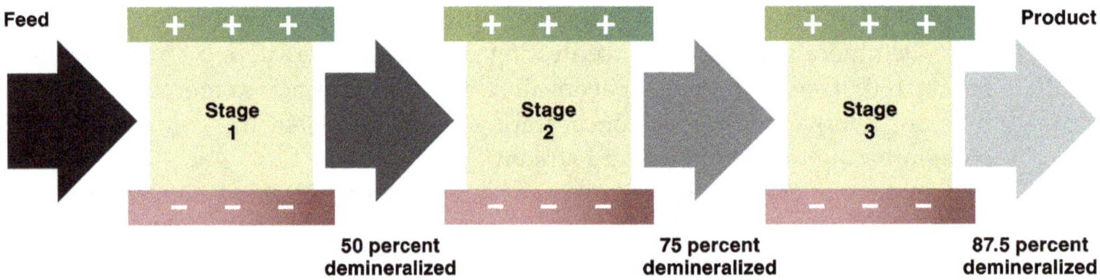

Adapted with permission from Ionics Inc.

Figure 3-16 An ED system with three hydraulic and three electrical stages

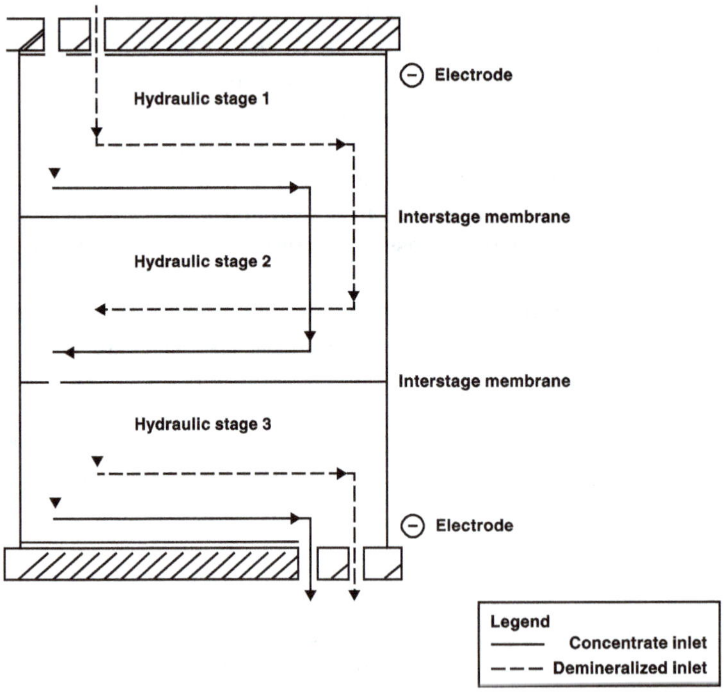

Adapted with permission from Ionics Inc.

Figure 3-17 Hydraulic staging within a single membrane stack

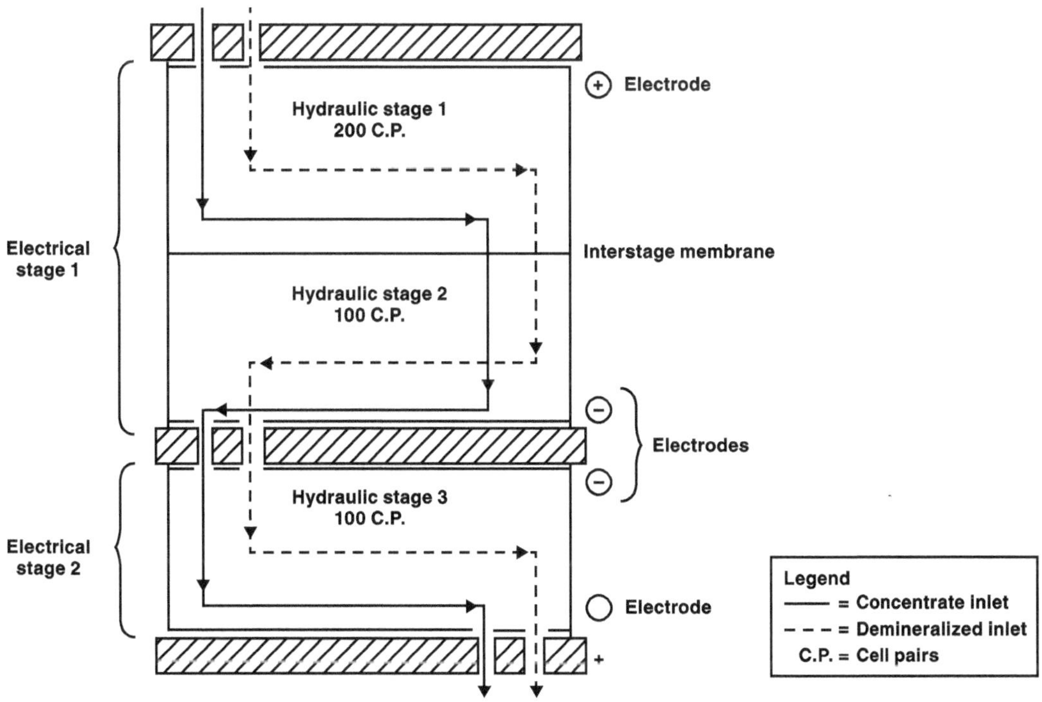

Adapted with permission from Ionics Inc.

Figure 3-18 Membrane stack with two electrical stages and three hydraulic stages

this section). Electrical staging applies independently controlled current to each electrical stage to remove a certain amount of salt from the source water. Figure 3-18 illustrates an example of electrical staging.

Current Density and Polarization

Current density can be described as an electrical flux expressed in terms of amperes per square centimeter. It applies to the mobility of individual ions, and the limiting factor is the rate at which ions can freely move through a solution. Ions moving through the solution should balance ions moving through the membrane. Current density is a function of fluid velocity in the flow path, stream temperature, and ionic composition.

As current density increases, ions in a solution migrate through the membrane faster than they move through the solution. The solution can become so depleted of ions that the water molecules are affected. Increased voltage causes water to dissociate into hydrogen ions (H^+) and hydroxyl ions (OH^-). This is called the polarization point, and it occurs in the demineralizing compartments. When polarization becomes pronounced, the pH of the demineralized stream changes. Hydrogen ions and hydroxyl ions migrate through the cation and anion exchange membranes, respectively. In the concentrate stream, OH^- tends to combine with opposite charged ions, such as magnesium. The resulting scale limits efficiency and requires frequent cleaning to prevent irreversible fouling in ED/EDR systems.

Operating at current densities sufficient to cause polarization results in the following consequences in ED/EDR systems:

- Energy consumption increases as electrical resistance increases.
- Current efficiency decreases because the transfer of hydrogen or hydroxyl ions is not resulting in the desired product water quality.

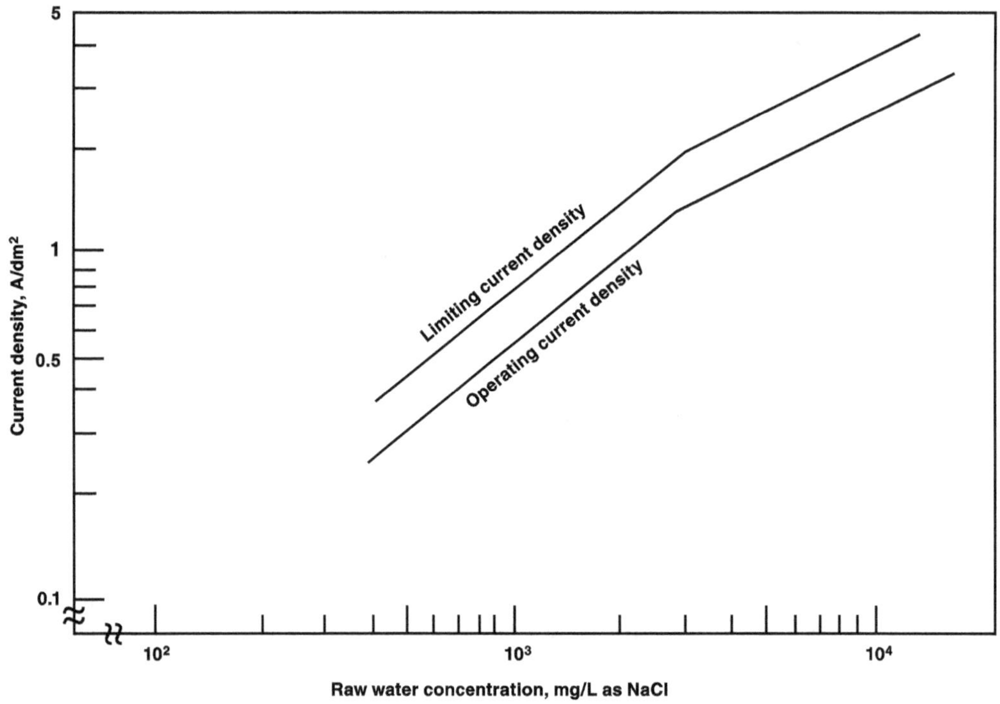

Adapted with permission from Hamada 1993

Figure 3-19 Raw water concentration and current density

- Additional increase of current becomes less efficient in transferring ions in polarization areas.

In commercial ED and EDR systems, a value of 70 percent of the limiting current density is used as the maximum allowable for brackish water operation, as illustrated in Figure 3-19. This provides a reasonable level of safety in system design.

Salt Removal Rate by Electrodialysis Reversal

When applying EDR for brackish water desalination, the two main approaches for reducing treatment costs are:

- reduction of capital cost by reduction of the size of EDR stacks, and
- reduction of the volume of concentrate to allow reduction of the cost of concentrate management.

In both areas, there are some niches in which the process improvements would result in lower overall cost of brackish desalination.

Sizing EDR equipment depends on several factors, such as feed flow rate, feedwater chemistry, criteria for the treated product water, and the flux rate of ions through ion-exchange membranes in the EDR stack. Feed flow rate, feed chemistry, and product requirements are the factors dictated by the treatment plant; the flux rate of ions through membrane, however, can be modified by design engineers. Increasing the flux rate by increasing the applied energy (i.e., higher applied voltage and, as a result, higher current) corresponds to higher rate of salt removal. To reduce the capital cost of an EDR plant, salt removal rate (flux rate of ions through membrane surface) should be maximized.

Understanding the limitations in increasing the flux rate helps the design engineers to reduce the membrane surface area and, as a result, reduce the capital cost of the EDR plant. As applied voltage to the EDR stack increases, current perpendicular to the membrane surface also increases and more ions are transferred from the boundary layer on the surface of the membrane in the dilute stream to the boundary layer on the surface of the membrane in the concentrate stream. However, increasing the voltage and therefore increasing the current density depletes the concentration of salt ions near the membrane surface, and it becomes progressively more difficult to remove additional salt. Ultimately, at a certain current density, the concentration of ions adjacent to the surface of the membrane in the dilute side becomes zero. This certain current density is called limiting current density and is one of the most important criteria in designing EDR systems. Reaching limiting current density has a number of adverse effects on salt removal: water dissociation into H+ and OH–, pH interruption, possible membrane damage, very high electrical resistance of the boundary layer and, as a result, high energy consumption. Factors such as hydrodynamics of the flow (e.g., flow rate), temperature, feedwater composition, and the shape and length of EDR stack determine limiting current density in an EDR stack.

Design engineers always try to maintain a safety margin to avoid nearing the limiting current density. However, the safety margin is not well defined, and the common safety margins used to design EDR systems might be too conservative. It appears that by increasing the conductivity cut rate per stage, the number of stages to achieve a target overall cut rate is reduced. This is a significant capital cost savings on the EDR desalination process. A better understanding of limiting current density and the safety margin will help design engineers to design EDR systems in a more efficient and aggressive manner, which would result in lower desalination cost.

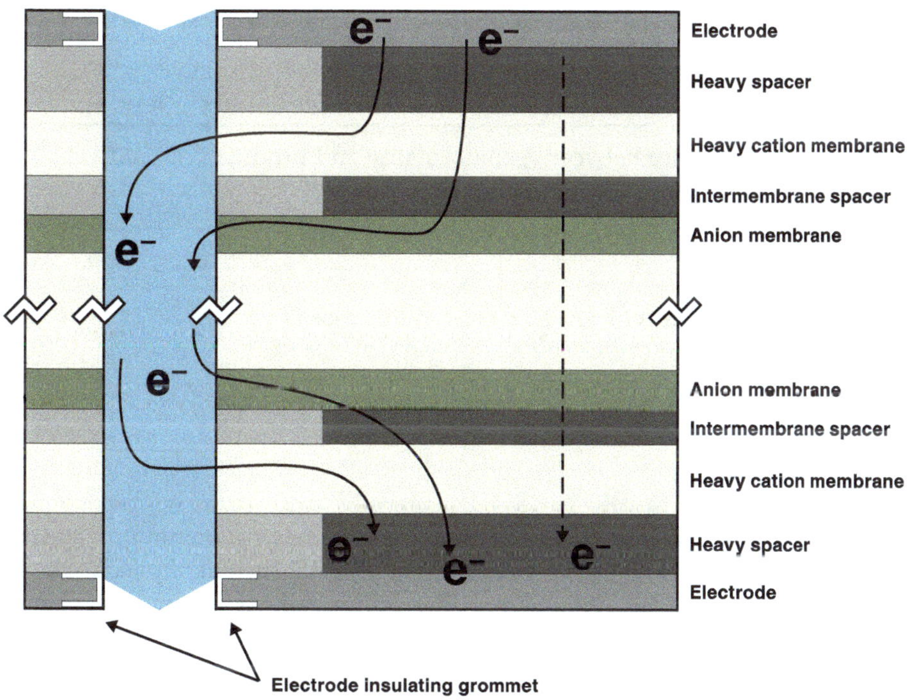

Adapted with permission from Ionics Inc.

Figure 3-20 EDR stack shorting

Current Leakage

Current leakage is defined as excess electric current at the electrodes that travels through an adjacent membrane, either a heavy anion or cation membrane, into higher conductivity water, usually the concentrate. Figure 3-20 illustrates stack shorting or current leakage, with the electrodes moving into the concentrate and then back toward the opposite electrode. The dotted line represents the normal movement of an electrode through the membranes and spaces without current leakage. Excess current can melt or "burn" the membranes and spacers. Normal design practices limit this voltage to 80 percent of the current that would cause burning. The limit is determined by water temperature, source water concentration, membrane stack size, and the internal manifold that splits flow into concentrate and dilute streams.

Electrical Current Efficiency

Current efficiency refers to the percentage of total current that is effectively transferring ions. In ED and EDR applications, the current efficiency e for the transfer of salts can be calculated based on the expressions in Faraday's law as follows:

$$e = \frac{F^* \times Fd \times \delta N \times 100}{I \times N^*} \qquad \text{(Eq 3-2)}$$

As an example, consider the following sample operating readings:

Source water (normality)	0.0480N
Product water (normality)	0.0078N
Demineralized flow rate (Fd)	9,765 L/h (43.0 gpm)
Current (I)	25.5 A
Number of cell pairs (N^*)	450
Faraday's constant (F^*)	26.8 A h (96,500 A s)

The difference between feed normality and product normality is

$$\Delta N = (0.0480 - 0.0078) = 0.0402$$

Therefore, the current efficiency in this case is

$$e = \frac{26.8 \times 9{,}765 \times 0.0402 \times 100}{25.5 \times 450} = 91.7 \text{ percent}$$

Commercial membrane stacks normally have a minimum current efficiency of 90 percent, which means that 90 percent of the total current applied to the membrane stack is effectively transferring ions from the demineralized stream to the concentrate stream. The remaining 10 percent of the current can be attributed to loss resulting from cross leakage and back diffusion.

Recovery Rates

The water recovery is expressed as a percentage and is defined as the total amount of water produced from the total amount of water supplied. EDR systems offer design options that optimize recovery, as discussed in this section. Off-specification product recycle, phased reversal, increased reversal cycle times, or any combination of these functions can be incorporated in EDR systems to take advantage of a substantial increase in recovery and production rates at minimal cost. Chemical dosing can further increase recovery rates and reduce waste disposal costs.

Concentrate Recycle

Recirculating the concentrate stream is called concentrate recycle. Flow rates of concentrate and demineralized water through the membrane stacks are essentially equal. Without recycling, this equal flow would lead to a recovery of demineralized water equal to half of the saline water treated. However, in most ED and EDR systems, the concentrated effluent is recycled to conserve water. When the recycle stream is eventually sent to waste, the concentration level of effluent is controlled by adding an equal volume of source water according to concentrate disposal regulations.

Off-specification Product Recycle

When the polarity reverses in an EDR system, the concentrate compartments become demineralizing compartments, and vice versa. The salinity in what used to be a concentrate compartment changes from a high TDS level to the product TDS level. The time needed for this change to take place is equivalent to the time required for source water entering this compartment to pass through the entire membrane system, normally 30 s per hydraulic stage. Product water leaving the system at this time is normally diverted to waste because it does not meet specifications. However, if reversal of the outlet valves is delayed, the average salinity of the product water can still be less than the salinity of source water. This product water can then be recycled back to the system feed, eliminating the need to send it to waste. This process is known as off-specification product recycle (OSPR).

Phased Reversal

Phased reversal (PR) in EDR is based on having independent control of:

- reversal of the inlet valves,
- the electrical polarity of each stage (electrodes), and
- reversal of the outlet valves.

PR works as follows: Suppose there are three stages in series. If the polarity was reversed on all three stages at once, the water contained in those stages at that instant would not meet product specifications when it flowed out of the final stage. In other words, three stages' worth of water would not meet product specifications. However, if the polarity was reversed on the second stage a certain amount of time *after* reversal on the first stage (to allow all of the off-specification water to finish passing through the first stage), only one stage's worth of water would not meet specifications. The timing of the third stage's polarity reversal would be handled similarly. Thus, the total reversal water is contained in a single stage in the system instead of through the entire system. PR increases recovery by increasing the amount of demineralized water produced while keeping the total amount of water consumed constant. The volume of waste during PR varies from 2 to 4 percent of the product water volume.

Reversal Frequency

In some cases, membrane scaling in an EDR system can be effectively controlled even with a reduced polarity reversal frequency. Standard systems use 15 min/polarity cycle, but some systems can operate with a 30-min cycle. Cutting the frequency of reversal in half reduces the portion of off-specification water by 50 percent. This increased net production augments recovery while reducing power consumption.

Electrode Stream Recycle

When cathode and anode streams are combined, they chemically neutralize one another and have a pH and salinity close to source water. Gases are collected and vented off, and liquid is normally sent to waste. Because this electrode stream quality is similar to that of source water, it may be recycled through the feed tank to increase system recovery and reduce total wastewater.

Chemical Addition

In EDR systems that do not use chemical dosing to prevent scaling, the maximum allowable concentrate salinity is determined by the calcium sulfate and calcium carbonate concentrations. If higher recoveries are desired, then chemical dosing is needed to combat scaling.

As necessary, an antiscalant can be added to control calcium sulfate scaling, and acid can be used to control calcium carbonate scaling. Only a small quantity of antiscalant (1–5 mg/L) is required because the streams are recycled. The saturation of calcium sulfate can then be increased from 150 percent of its value in the absence of chemical addition to 275 percent. Acid is required to maintain the LSI at +1.8 in the concentrate stream for calcium carbonate control.

Chemical Cleaning

Clean-in-place (CIP) process. An ED or EDR unit is equipped with a CIP system to allow periodic flushing of the membrane stack and piping with an acid solution. A chemical feed pump and storage tank form the main components of the CIP system. The ED/EDR CIP system is similar to the RO CIP system and contains the following components:

- booster pump (smaller systems use the concentrate pump on skid),
- CIP tank with heater,
- chemical feed pump or drum pump to add cleaning chemical to tank,
- flowmeter, and
- cartridge filter (optional).

The CIP process is intended to remove scale or reversible fouling from the membrane stack and hydraulic piping. The chemical solution circulated through the stack depends on the type of contamination. This is a required procedure used in preventive maintenance that should be completed according to manufacturer's recommendations.

Only the chemical solutions approved by the membrane manufacturer should be used in the CIP process to avoid irreversible damage to the membrane. The pH limits as provided by the membrane supplier must be strictly followed. The following chemical solutions are commonly used.

Hydrochloric acid (HCl) solution. Periodic cleaning with a 2–5 percent HCl solution is the most frequently used method. The solution is used to remove carbonate scale as well as some metal deposits and light dirt. The procedure requires the use of the CIP tank, deionized water, and hydrochloric acid.

Sodium chloride (NaCl) solution. A 3–5 percent NaCl solution removes organic foulants, which are present in many surface waters, from the membranes. This procedure requires the CIP tank, deionized water, sodium chloride (food grade), sodium hydroxide (NaOH), and pH test paper (pH 8.0–10.0).

The solution should be at least 3 percent NaCl (0.55*N* chloride) and have a pH approaching the maximum pH allowed by the membrane supplier. There are some anion membranes that would be damaged by the use of pH greater than 11. This solution should then be circulated through the system. After the NaCl application, the operator should flush the membranes with HCl to remove excess salt and any other precipitates that may have formed because of the high pH operation.

Chlorine solution. A 10- to 50-mg/L chlorine solution disinfects the membranes and hydraulic piping. This procedure requires the CIP tank, deionized water, and 5.25 percent commercial hypochlorite solution mixed at a rate of 0.25 gal (1 L) hypochlorite per 500 gal (1,890 L) water. Some membranes can be damaged by the application of chlorine; the membrane supplier should be consulted for the recommended disinfection procedure.

Stack Repair and Disassembly

The membrane stack may be removed from the ED or EDR unit if stack repairs are necessary. Repairs can be completed manually with the assistance of an overhead crane. Stacks should not be allowed to remain dry for long periods of time (i.e., more than 24 hours) because membranes may become damaged. Blind flange connections should be attached to the remaining stacks to prevent them from drying. Specific manufacturer's instructions for stack removal should be followed.

When a problem arises, stacks can be disassembled in accordance with procedures recommended by the system's manufacturer. Generally, disassembly requires that each piece be removed separately, with the exception of the top electrode, which can be replaced without removing any of the membranes.

It is important to maintain correct component orientation and to store membranes in water. The stack should be rebuilt in the order it was disassembled; incorrect assembly can impede performance or cause scaling.

During normal operation of an ED or EDR plant, chemicals used in operation, cleaning, and maintenance include:

- sodium hexametaphosphate or other antiscalants, for scaling prevention;
- chlorine, for disinfection or oxidation of arsenic in feedwater;
- acid solution, for cleaning and scale removal in the stacks and electrode compartments;
- salts, for cleaning organic deposits from membranes; and
- caustic solution, for cleaning organic deposits from membranes.

Chemicals should be placed in a chemical storage room according to storage directions, and all safety precautions should be observed when chemicals are being handled.

Chemical room design is very important and requires careful review. As HCl is a common component in EDR systems, the chemical room should be designed with the fuming of this chemical in mind. Additionally, acid and chlorine should never be stored near each other.

Safety Considerations

Proper operation of an ED or EDR system requires an understanding of the safety issues involved in electrical operation, chemical handling, and mechanical systems.

Electrical Operation

There are many factors relating to electrical operation of an ED or EDR system that must be considered.

Equipment shutdown. During normal operation, an operator can shut down the ED or EDR unit by placing the system in manual and pressing the stop-start button. The unit will then enter a flush cycle. When this cycle is completed, the unit will be idle. In an emergency, however, the unit should be turned off at the main circuit breaker because the stop-start button will begin the flush cycle.

Grounding. All ED and EDR systems, including the stacks, should be connected to an electrical ground of earth potential to prevent electric shock from stray voltages. If the unit is skid mounted, a separate earth ground should be attached to the skid. Stacks should then be grounded by a wire to the skid or electrical cabinet. Grounding cables are provided by the manufacturer and should be installed so that they are not easily removed.

When equipment is moved, dismantled, or repaired, it is important to make sure that all electrical ground connections are restored.

High voltage. ED and EDR equipment use high-voltage AC power. This AC voltage is converted to DC by a rectifier for use in the operation of the membrane stacks. DC voltage should never be applied to the stacks unless water is flowing through the stacks.

The connecting bolts for the electrode tab should be tight and corrosion free. Overheating due to loose connections could damage the membrane stack.

DC amperage will rise and fall according to source water temperatures. To save power and prevent damage to the stack when source water temperatures rise, DC stack voltages must be lowered until DC amperage returns to normal settings. The DC stack voltage should not be increased as source water temperatures drop unless the manufacturer provides specific instructions to do so.

Membrane stack. It is important not to touch the wet stack sides, electrode plates, or the electrode plate tab connections when the unit is operating. Operators should always assume that the stacks are electrically alive. Protective clothing, such as rubber gloves, should always be worn during a stack probe. The equipment should not be washed down when the power is on.

Electrical panels. All control panels contain live voltage. Some systems are designed so that the main power must be off before the doors can be opened. Extreme care must be taken to avoid contact with electrically live components.

Only staff experienced in servicing the control panel should troubleshoot the panel. Equipment should never be operated with control doors open unless the cabinet is being repaired. This will protect personnel from electrical hazards and will also protect the electrical system from dust and water.

Chemical Safety

The degasifier tank separates the dissolved gases from the electrode waste streams. The system should be vented to the atmosphere at least 1 ft (0.3 m) above the highest peak of the building. Personnel should always wear protective clothing (rubber gloves, apron, face shield, goggles, boots, and chemical-protection suit) when handling corrosive solutions or when working on the chemical pumps and supply lines. In addition, they should always wash down the work area after handling the chemicals and always clean up any chemical spills. When acid and water are being mixed, it is important to *always* add the acid to the water. Adding water to acid can cause a violent explosion, spraying acid on the operator.

Table 3-3 Typical water impurities and corresponding pretreatments

Impurity	Pretreatment	ED–EDR Tolerance Limit
Turbidity	Filtration	2 NTU
Iron	Oxidation and/or filtration	0.3 mg/L
Manganese	Oxidation and/or filtration	0.1 mg/L
Hydrogen sulfide	Oxidation	0.3 mg/L
Chlorine	Carbon adsorption or sodium bisulfate	0.5 mg/L (continuous)

Source: Ionics Inc.
ED—electrodialysis, EDR—electrodialysis reversal, NTU—nephelometric turbidity units.

Mechanical Systems

It is necessary to remove the stack end plates when stack maintenance is being performed. A hoist should be used since the plate is very heavy. It is important to always move the plate out of the way and set it down on a firm surface before proceeding.

A centrifugal pump should never be operated with either the discharge or suction valve closed. This can cause friction to the impeller and heat the trapped volume of water to a high temperature. If this heated water escapes, it can cause serious burns.

It is also important never to operate any rotating equipment (specifically pumps, blowers, or fans) without their protective guards in place.

ELECTRODIALYSIS/ELECTRODIALYSIS REVERSAL PRETREATMENT

ED may require some form of source water pretreatment for successful, long-term operation. Minimal pretreatment includes suspended solids removal by 10-m cartridge filtration. Additional pretreatment may be required, such as conventional coagulation, sedimentation, and filtration, depending on source water quality. ED and EDR normally require less pretreatment than NF/RO membrane processes because the flow path for the water is larger. Table 3-3 summarizes pretreatment methods for typical impurities encountered in source waters.

Turbidity

The efficient operation of ED and EDR requires a source water turbidity level of no more than 2 NTU. Because surface water turbidities can vary from 2 to 2,000 NTU, pretreatment may be required. Sand or multimedia filtration may be necessary to achieve consistently low suspended solids or turbidity levels. Highly turbid sources may require in-line chemical coagulation and clarification followed by filtration.

Most groundwater sources are consistently free of organic or suspended turbidity particles, making them suitable for direct ED and EDR treatment. The only pretreatment normally used with a groundwater source is an in-line 10-m cartridge filter that is supplied as part of an ED or EDR system.

Iron and Manganese

Iron (Fe) and manganese (Mn) are most frequently found in groundwater from deep wells. Their concentrations often range from 0 to 15 mg/L. Maximum levels tolerated by ED and EDR are 0.3 mg/L Fe and 0.1 mg/L Mn. Pretreatment for iron and manganese includes aeration and filtration, chemical oxidation followed by filtration, or a form of oxidizing filtration such as manganese zeolite filtration.

Hydrogen Sulfide

A gas detectable by its characteristic "rotten egg" odor, hydrogen sulfide (H_2S) concentrations can range from 0 to 50 mg/L in groundwater. The ED and EDR tolerance limit is 0.3 mg/L, with the same recommended pretreatment as that for iron and manganese removal.

Chlorine

Chlorine may be present in some source waters as residual from upstream bacterial control. Ion-exchange membrane tolerance is 0.5 mg/L for free chlorine on a continuous basis. Membranes can be disinfected with a short-term dosage of 50 mg/L chlorine when necessary for bacterial control. Overall chlorine tolerance varies with membrane material and manufacturer. Typical pretreatment for chlorine removal is with carbon adsorption or sodium bisulfate.

Scale

The recovery rate, or the percentage of process source water that is recovered as treated product water, can be limited by the presence of certain salts in the concentrate stream that can form a scale on the membrane surface. As water passes through the ED or EDR process, source water salts are removed and concentrated in the concentrate stream. The portion of concentrate that is wasted determines the concentration of salts in the concentrate or concentrate stream and the rate of recovery. In other words, repeatedly recycling the concentrate stream before sending it to waste creates a high concentration of salts in the concentrate stream and increases the recovery percentage. High water recovery minimizes wastewater pretreatment, reduces pumping costs, and may decrease waste disposal costs. A high concentration of salts, however, increases scaling potential if the concentration is beyond solubility limits. Calcium carbonate and sulfate salts of calcium, barium, fluoride, and strontium are the most common scale formers. The potential for scale formation or membrane fouling also differs with the type of ED used.

Electrodialysis Systems

ED plants have strict limitations on concentrate chemistry and may require presoftened source water to limit calcium sulfate scale formation. The concentrate stream for ED must have an LSI of zero, meaning the water would neither dissolve nor precipitate calcium carbonate. (See Chapter 4 for additional information.) ED systems can tolerate up to 90 percent calcium sulfate saturation without the need for an antiscalant. Higher levels of calcium sulfate saturation, up to 150 percent and an LSI of 1.5, are acceptable for an ED system with an antiscalant such as sodium hexametaphosphate maintained in the concentrate stream. Because the concentrate stream is recycled, only a low antiscalant feed is needed to maintain a residual of less than 10 mg/L.

ED systems can use specially impregnated membranes to remove specific ions (e.g., monovalent membranes to remove sodium chloride) and divalent ions to remove nitrate. ED systems can also make use of two hydraulic stages, with anion monovalent permselective membranes in the first stage and cation monovalent permselective membranes in the second stage. By keeping the concentrate streams from the two stages separate (and thus keeping the calcium and sulfate ions separate within the ED unit), the system can often reach effective calcium sulfate levels that are many times the normal saturation point when it finally mixes the two concentrate streams outside of the ED unit itself. Levels two to six times greater than calcium sulfate saturation can often be achieved, depending on the specific chemistry. However, at no time should the calcium sulfate concentration in a specific stage exceed 90 percent or 150 percent of saturation, depending on the absence or presence of scale inhibitors. Overall water recovery rates for such systems can reach as high as 95 percent on brackish waters.

Electrodialysis Reversal Systems

The polarity-reversing EDR system requires less pretreatment than ED and is capable of higher water recovery. EDR's regular and automatic reversal of the DC polarity drives membrane foulants and scale deposits from membranes into the concentrate stream and out of the system. Without chemical antiscalants, the EDR concentrate stream can handle an LSI as high as +1.8 and a calcium sulfate saturation of 150 percent. Recovery rates can be further enhanced by adding antiscalant to the concentrate stream to achieve an LSI of +2.5 and a calcium sulfate saturation of 275 percent. This enables typical EDR systems to achieve water recovery rates from 80 percent to more than 90 percent.

ED and EDR membrane systems can be cleaned in place to restore system performance after scale formation or reversible membrane fouling. Scale is removed by circulating an up to 5 percent acid solution, usually hydrochloric acid, through the membrane stacks. Reversible organic fouling is removed by circulating a 5 percent brine solution with a pH of 10 through the membrane stack. Chlorine can be added to the brine solution at a concentration of 10—50 mg/L to disinfect membranes and aid removal of organic matter. Severe fouling or scale can be remedied by manually cleaning a disassembled membrane stack, resulting in little or no loss in membrane use or operating life.

ELECTRODIALYSIS/ELECTRODIALYSIS REVERSAL POST-TREATMENT

Post-treatment requirements for ED and EDR systems depend on the intended use of the product water (i.e., industrial or potable). Each application has different post-treatment needs.

Industrial users generally have water purity specifications that exceed drinking water standards. In ED and EDR systems, the only post-treatment normally found is ion exchange. Mixed-bed ion exchange units remove the remaining ions from the ED and EDR product water, serving as "polishers" in these high-purity applications.

In municipal or potable water applications, product water must meet both primary and secondary regulations established by the SDWA. In addition, post-treatment is often required for disinfection and corrosion control.

Potable ED and EDR plants are usually equipped with a small chlorinator to provide the level of disinfection required by federal and local health agencies. Chlorine is used most frequently to reduce pathogenic microorganisms, and ozone is used for disinfection in some bottling applications.

Because of the corrosive nature of membrane-treated water, a stabilizing agent may be added to prevent or minimize oxidation of metallic surfaces. The most common agent is an alkali, such as caustic soda (NaOH) or ammonium hydroxide (NH_4OH), but it may also be a metal complex, such as zinc phosphate [$Zn_3(PO_4)_2$]. The metallic phosphate forms a barrier between the water and any metal surface, providing a protective coating. The treatment objective is to maintain the LSI around zero or slightly positive to prevent pipe corrosion.

REFERENCES

Adham, S.S.; Trussell, R.S.; Gagliardo, P.F. ; and Trussell, R.R. 1998. Rejection of MS-2 Virus by RO Membranes. *Journal AWWA*, 90(9):130–135.

Andrews, L.S. 1991. Disposal Methods and Issues of Concentrate Discharges—An Overview. Proc. AWWA Membrane Conf., Baltimore, Md.

AWWA. 2014 (2nd ed.). M38, *Electrodialysis and Electrodialysis Reversal*. Denver: AWWA.

AWWA. 2007 (1st ed.). M38, *Electrodialysis and Electrodialysis Reversal*. Denver: AWWA.

Carpenter, F. and Geishecker, E. 1989. A Quarter Century of Electrodialysis Desalting: The Buckeye, Arizona, Experience. Proc. 108th Annual Conf. of AWWA, Los Angeles, Calif.

Conlon, W.J. 1988. Disposal of Concentrate From Membrane Process Plants. Proc. AWWA Membrane Conf., Orlando, Fla.

CRC. 2005. Ionic Conductivity and Diffusion at Infinite Dilution. In *Handbook of Chemistry and Physics* (P. Vanýsek, editor). Cleveland, Ohio: CRC Press/Taylor and Francis.

Frenkel, V. 2009. Membrane Technologies for Wastewater Treatment and Water Reuse. 2009 Water Environment Federation Annual Technical Exhibition and Conf., Orlando, Fla.

Frenkel, V. 2007. Brackish vs. Seawater Desalination: Which One is for You? Proc. 2007 IDA World Congress on Desalination and Water Reuse, Maspalomas, Gran Canaria, Spain.

Gagliardo, P.F.; Adham, S.S.; Trussell, R.R.; and Olivieri, A.W. 1997. Membranes as an Alternative to Disinfection. Proc. 1997 AWWA Annual Conf. and Exhibition, Atlanta, Ga.

Hamada, M. 1993. Brackish Water Desalination by Electrodialysis. *International Desalination and Water Reuse Quarterly*, 2(4):8–15.

Helfferich, F.G. 1962. *Ion Exchange, Advanced Chemistry Series*. New York: McGraw-Hill.

Inoue, S. and Kuroda, O. 1993. Electrodialysis Desalination System Power By Photovoltaic Power Generation. *Industrial Application of Ion Exchange Membranes*, 2:177–180.

Kemperman, A.J.B. 2000. *Handbook on Bipolar Membrane Technology*. Enschede, The Netherlands: Twente University Press.

Loeb, S. and Johnson, J.S. 1967. Operating a Reverse Osmosis Desalination Plant. *Chemical Engineering Progress*, 63(1):90.

Mallevialle, J.; Odendaal, P.E.; and Wiesner, M.R. 1996. *Water Treatment Membrane Processes*. New York: McGraw-Hill.

Mi, B.; Eaton, C.L.; Kim, J.; Colvin, C.K.; Lozier, J.C.; and Marinas, B.J. 2004. Removal of Biological and Non-Biological Viral Surrogates by Spiral-Wound Reverse Osmosis Membrane Elements With Intact and Compromised Integrity. *Water Research*, 38(18):3821–3832.

Mintz, M.S. 1963. Electrodialysis—Principles of Process Design. *Industrial and Engineering Chemistry*, 55(6):18–28.

Moscovis, V. 2013. The Outcomes of Perth's Groundwater Replenishment Trial. Proc. Australian Water Association Conf., Perth, Western Australia.

Newman, J.S. and K.E. Thomas-Alyea. 2004 (3rd ed.). *Electrochemical Systems*. Hoboken, N.J.: John Wiley & Sons.

Pontius, N. 1993. *Back to Basics Guide to Corrosion Control for Lead and Copper*. Denver: AWWA.

Pype, M.; Patureau, D.; Wery, N.; Poussade, Y.; and Gernjak, W.D. 2013. Monitoring Reverse Osmosis Performance: Conductivity Versus Fluorescence Excitation-Emission Matrix (EEM). *Journal of Membrane Science*, 428:205–211.

Robinson, R.A., and Stokes, R.H. 1959. *Electrolyte Solutions—The Measurement and Interpretation of Conductance, Chemical Potential and Diffusion in Solutions of Simple Electrolytes*. London: Butterworths Scientific Publications.

Rynders, T.; Stillwell, C.; and Niebur, S. 2013. The Heat is On: Startup at ECCV's 10.9 mgd Inland RO Facility During Record Breaking Temperatures and Demands. Proc. AWWA/AMTA Membrane Technology Conf., San Antonio, Tex.

Shaw, M.; LeTourneau, T.; Liang, L.; and Ng, K.H. 2011. Low Energy Desalination: Siemens' Solution to a Global Challenge. IDA World Congress, Perth, Australia.

Siwak, L.R. 1993. Here's How Electrodialysis Reverses and Why EDR Works. *International Desalination and Water Reuse Quarterly*, 2(4):16.

Solan, A.; Winograd, Y.; and Katz, U. 1971. An Analytical Model for Mass Transfer in an Electrodialysis Cell With Spacer of Finite Mesh. *Desalination*, 9(1):89–95.

Sonin, A. and Probstein, R.F. 1968. A Hydrodynamic Theory of Desalination by Electrodialysis. *Desalination*, 5(3):293–329.

Strathmann, H. 2004. *Ion-Exchange Membrane Separation Processes*. Amsterdam: Elsevier.

Tanaka, Y. 2007. *Ion Exchange Membranes: Fundamentals and Applications*. Amsterdam: Elsevier.

Veza, J.M.; Penate, B.; and Castellano, F. 2004. Electrodialysis Desalination Designed for Off-Grid Wind Energy. *Desalination*, 160(3):211–221.

Wetterau, G.D.; Chalmers, R.B.; Liu, P.; and Pearce, W. 2013. Advancing Indirect Potable Reuse in California. *Water Practice & Technology*, 8(2):275–285.

Wetterau, G.D.; Desormeaux, E.D.; and Luckenbach, H. 2011. Achieving Regulatory Acceptance of Seawater Treatment. Proc. 2011 AWWA Annual Conference and Exhibition, Washington.

Wolfe, T.D.; Bourcier, W.L.; and Metcalfe, P. 2011. Osmotic Pressure Calculation Using Pitzer Equation Modeling. Proc. 2011 IDA World Congress, Perth, Australia.

This page intentionally blank.

Chapter 4

Discharge Options for Concentrate Disposal

Rick Bond and Sandeep Sethi

In this manual, discharge options are defined as those that do not involve treatment to recover all or part of the concentrate as product water. Discharge options for concentrate disposal are generally operationally simpler and less expensive than options involving concentrate treatment, but they offer less opportunity for water conservation and sustainable management of salinity.

The discharge methods described in this chapter include:

- discharge to surface water,
- discharge to sewer or brine-line,
- deep-well injection,
- evaporation ponds,
- enhanced evaporation systems, and
- irrigation.

Historically, most desalting facilities have disposed of concentrate by discharge to surface water, discharge to sewer, or deep-well injection. In a 2003 survey of concentrate management practices in the United States, 85 percent of the 234 municipal desalting plants in operation used discharge to surface water, discharge to sewer, or deep-well injection for concentrate disposal (USBR 2006).

The 2003 survey also revealed trends in the type of concentrate discharge option used based on plant size.

- Surface water discharge was used by plants of all sizes.
- Sewer discharge was used less frequently as desalting plant size increased because of the impact of large concentrate flows on the receiving wastewater treatment plant or water resource recovery facility (WRRF).
- Deep-well injection was used less frequently by small plants because economies of scale for well construction favor high-capacity wells.
- Evaporation ponds and irrigation were used only by smaller plants because of the extensive land-area requirements.
- No enhanced evaporation systems were reported.

Geography and climate also influence the method of concentrate discharge selected. Discharge to surface water is generally not an option in the arid southwestern United States where surface water is scarce and highly valued. There has been a trend for municipalities in water-scarce regions to make greater use of reclaimed water, and quality requirements for specific uses of reclaimed water can limit discharge of concentrate to the sewer. Evaporation ponds are better suited to level land and a warm, dry climate. The ability to irrigate is influenced by climate, particularly precipitation rates.

Each of the discharge methods is discussed in the following sections.

DISCHARGE TO SURFACE WATER

Historically, discharge to surface water has been the most widely used concentrate management method (USBR 2006). Surface water discharge generally has low cost but high standards for environmental protection relative to other options. Environmental restrictions increasingly limit the use of surface water discharge for concentrate disposal. Since the early 1990s, only five desalination facilities in Florida have selected surface water discharge while 34 selected deep-well injection (Robert 2010). To protect water resources from degradation, current environmental regulations essentially limit discharge of concentrate to surface water of similar salinity. For example, California does not allow discharge of concentrate to a receiving surface water of lower salinity. Consequently, surface water discharge is not an available option for many inland desalination plants, particularly for those in the arid Southwest where surface water sources are scarce.

Regulations Governing Discharge to Surface Water

In the United States, discharge of concentrate to surface water is regulated under the federal Clean Water Act (CWA) through the National Pollutant Discharge Elimination System (NPDES). NPDES was created under the CWA to establish minimum standards for surface water discharge, and any direct discharge to waters of the United States requires an NPDES permit. NPDES permits specify effluent limits, pretreatment requirements, and standards for inspection, monitoring, and reporting.

The CWA required every state to develop water quality standards applicable to all water bodies or segments of water bodies within the state. Each water body is classified by its designated uses, and NPDES limits are designed to ensure that water quality is preserved for those uses. In general, NPDES permit limits are structured to ensure that the water is safe for aquatic life, drinking water use, and public health exposure. For many noncarcinogenic contaminants, standards designed to protect aquatic life are more stringent than drinking water standards because aquatic life is more sensitive than humans to the contaminant.

The USEPA uses an integrated approach for preventing discharge of toxic substances to surface waters. This approach starts with defining water quality objectives, criteria, and standards. Next, an NPDES permit is developed that sets limits designed to control toxic pollutants through both chemical-specific and whole effluent toxicity (WET) limits. WET tests are conducted to evaluate the effect of a concentrate discharge on aquatic life. In WET tests, bioassay tests are conducted in which select test organisms are exposed to the concentrate and to dilutions of the concentrate for specified durations. Invertebrates, plants, and fish may be used in the test. Discharge concentrations and flow rates may also be limited by total maximum daily load (TMDL) studies. For a more detailed description of how NPDES criteria for surface water discharge are established, please see USEPA guidance manuals (USEPA 2010, 1991).

Design and Cost Considerations

The effect of concentrate discharge on receiving waters depends on the dilution and mixing capacity of the water. The amount of dilution and mixing required increases in proportion to the concentration of contaminants in the concentrate relative to the contaminant standard. Hence, environmentally safe surface water discharge becomes more challenging as the volume of the concentrate stream increases and as the salinity and concentration of contaminants of concern in concentrate increase.

Density of the concentrate relative to the receiving water also has an effect. If the concentrate has higher salinity than the receiving water, it will be denser and tend to sink. Conversely, if the concentrate has lower salinity, it will rise. Either condition could increase localized salinity gradients harmful to plants and fish.

Surface water discharge systems comprise (1) pumping and pipeline from the plant to the shore; (2) pipeline from the shore to the outfall; and (3) the outfall structure. Outfall structures frequently include a diffuser to promote dilution of the discharge. Diffuser design is complex and is often conducted using software programs designed for this purpose. Design variables include the following:

- discharge flow rate,
- shape and hydraulics of the receiving body,
- temperature and pH of the effluent and receiving body, and
- density of the effluent relative to the receiving body.

In some cases, the regulatory agency may grant an NPDES permit that allows for a mixing zone. A mixing zone is an allocated zone at the end of the discharge pipe where initial dilution occurs and contaminants are allowed to exceed water quality criteria as long as toxic conditions are avoided. Two mixing zones may be prescribed: an acute mixing zone and a chronic mixing zone. As the names imply, water quality criteria associated with acute health effects must be met in the acute zone and water quality criteria associated with chronic health effects must be met in the chronic zone.

Permits allowing mixing zones typically require mixing zone analyses with a steady-state model, a dye test, or both to compute dilution factors. Such analyses must be conducted for a variety of receiving water and concentrate discharge conditions to determine reasonable worst-case conditions for acute health effects and average conditions for chronic health effects.

Variables evaluated in a multivariable model to determine dilution factors include the following:

- concentrate flow rate;
- receiving water flow rate (for rivers and streams)—both low- and high-flow conditions may be required;
- depth of flow;

- stratification of the receiving water body;
- pollutants of concern in the concentrate;
- width of the receiving water body—some permits restrict the width of a water body that can be occupied by both acute and chronic mixing zones;
- concentrate hardness and pH; and
- temperature of the concentrate and the receiving water.

Water quality criteria for the analyses are selected based on potential health effects on humans and aquatic life for each of the contaminants of concern. In the CWA, the USEPA addresses the following three types of pollutants in effluent discharges:

- Toxic pollutants are the 126 pollutants specified in Section 307(a) of the CWA. These pollutants are sometimes referred to as priority pollutants. Concentration of substances during the desalination process may cause one or more of these pollutants to be a concern in RO concentrate.
- Conventional pollutants are defined by Section 304(a)(4) of the CWA as biochemical oxygen demand (BOD), total suspended solids (TSS), pH, fecal coliform, and oil and grease. These pollutants are typically not a concern in desalination concentrate.
- Nonconventional pollutants are any pollutant not already defined as a toxic or conventional pollutant. TDS fall under this category, and in many cases, high levels of TDS relative to the receiving water present the greatest permitting challenge.

The USEPA addresses three types of pollutant concentration in the CWA:

- Criterion maximum concentration (CMC) is the highest concentration of a pollutant that organisms can be exposed to for a brief time without causing an acute effect.
- Criterion continuous concentration (CCC) is the highest concentration of a pollutant that organisms can be exposed to indefinitely without causing unacceptable effect.
- Reference ambient concentration (RAC) is the concentration that will not cause adverse health effects in humans.

While surface water discharge is often the least expensive concentrate management option, there are many factors that affect its cost and feasibility. These include uses and water quality of the receiving water, distance to the receiving water, environmental impact, and outfall design. Furthermore, the NPDES permitting process can be lengthy and expensive.

Advantages and Disadvantages

Table 4-1 summarizes the advantages and disadvantages of discharge to surface water. The advantages of surface water discharge are that it is relatively inexpensive and requires little maintenance. The disadvantages of this option relate to the difficulties in finding an appropriate receiving water body and then obtaining an NPDES permit.

Table 4-1 Advantages and disadvantages of discharge of concentrate to surface water

Advantages	Disadvantages
• Typically inexpensive relative to other options • Minimal use of mechanical equipment; therefore, maintenance is relatively easy and inexpensive	• Potential for degradation of surface water quality • Requires an NPDES permit, and steps to acquire the permit may be lengthy, rigorous, and expensive • Limited by the availability of an appropriate receiving water • NPDES permit renewal necessary at the end of its term

DISCHARGE TO SEWER

Historically, discharge to sewer has been the secondmost widely used concentrate management method (USBR 2006). An NPDES permit is not required for discharge to the sanitary sewer system, but permission is required from the receiving wastewater treatment plant (i.e., a WRRF). All WRRFs have an NPDES discharge permit, and accepting desalination concentrate can diminish the ability of the receiving WRRF to comply with its NPDES permit limits. Moreover, concentrate may impair the ability of the WRRF to implement water reuse, which is the trend in recent years, particularly in drought-stricken areas.

High salinity levels can be toxic to the microbial population in a wastewater treatment plant; consequently, concentrate can adversely affect biological treatment. Furthermore, many inorganic constituents in the concentrate, such as chloride, are not removed in biological treatment; as a result, desalination concentrate may cause the WRRF to exceed its permit limits for inorganic contaminants. Concentrate can also adversely affect management of waste solids at the WRRF. Contaminants in concentrate that are removed by biological treatment end up in the waste sludge and diminish its suitability for land application.

Discharge to sewer is a simple and inexpensive option, but there are barriers to its implementation. WRRFs must be willing to accept the concentrate, and they typically impose local limits on concentrate water quality to protect treatment efficiency and ensure compliance with their NPDES permit limits. If the WRRF and desalination facility are not owned by the same entity, the WRRF typically imposes a fee to accept the concentrate. The fee may be based on flow, solids loading, or both. The amount of the fee is up to the WRRF, and such fees can be high. Agreements between desalting plants and WRRFs to accept concentrate typically have a set duration. At the end of the contract period, the WRRF can raise the discharge fee or decide it is no longer able to accept the concentrate. Consequently, once a desalination facility overcomes the initial hurdles, there is no guarantee that discharge to sewer will be a sustainable solution.

Factors that influence the ability to discharge to the sewer include the following:

- whether the desalination facility and the WRRF are owned by the same entity,
- relative flow rates of the desalination facility and the WRRF,
- TDS level of the concentrate as well as the concentration of specific inorganic substances that are not removed during treatment at the WRRF,
- potential impact on biological treatment at the WRRF,
- potential effect of contaminants in concentrate on WRRF sludge disposal,
- distance between the desalination facility and the WRRF,
- capacity in the sewer system between the desalination facility and the WRRF,
- future capacity increases by the desalination facility,
- allocation of existing or future WRRF capacity to planned service area expansion, and
- impact on the NPDES permit of the WRRF and on the reuse water, if applicable.

The CWA contains national water quality provisions for discharges to sanitary sewers under the national pretreatment program. These regulations are intended to protect publicly owned treatment works (POTWs) from high concentrations of industrial pollutants. There are two categories of standards: categorical pretreatment standards and specific prohibitions. Categorical standards are applied to specific wastewater discharges of particular industrial categories. POTWs are not included in any industrial category and therefore are covered under the specific prohibitions. Specific prohibitions are intended to

Table 4-2 Advantages and disadvantages of discharge of concentrate to sewer

Advantages	Disadvantages
• Typically inexpensive relative to other options • Minimal maintenance required	• Potential adverse impact on treatment at the WRRF and on NPDES permit • May require reliance on another entity (WRRF owner) • May not have future assurances of price and willingness of WRRF to accept concentrate

protect the receiving POTW from contaminants that could create a hazard, adversely affect treatment, cause corrosive damage, interfere with flow, or pose health risks to plant personnel. Several states have been delegated authority to administer the pretreatment program.

Table 4-2 summarizes the advantages and disadvantages of discharge to sewer. The biggest advantage of sewer discharge is that it can be inexpensive relative to other options. If the desalting facility and WRRF are owned by the same entity, then the owning entity has to be concerned with the effect of concentrate on treatment at the WRRF and the NPDES water quality limits for WRRF effluent. If the WRRF is owned by a separate entity, there are potential disadvantages associated with the desalting facility being reliant on another entity for its concentrate management solution.

Dedicated brine lines are a relatively new method of concentrate discharge and may be considered a subset of sewer discharge. The dedicated brine lines, however, discharge to the ocean rather than to a WRRF. The cost of dedicated brine lines is largely affected by distance from the ocean; consequently, this option is not a practical solution for inland desalination facilities other than those near the coast.

DEEP-WELL INJECTION

Deep-well injection is a disposal option in which waste streams are injected by a well into a porous subsurface formation for storage and confinement. Deep-well injection is a storage method for disposing of concentrate, and the intent is that the injected concentrate should remain immobilized in the storage zone indefinitely. Deep-well injection requires a suitable subsurface formation, so site selection is critical to the success of this discharge method. Extensive and expensive studies and testing are typically required to determine if a subsurface formation is appropriate for deep-well injection. This upfront work can include environmental, geological, and hydrological studies; drilling test holes; and drilling a test well. Such studies can involve considerable cost with no guarantee that the site will be determined to be viable.

In the United States, the majority of desalting plants using deep-well injection for concentrate disposal are in Florida, where hydrogeological conditions are suitable. Of the 26 plants in the United States using deep-well injection in 2003, 25 were in Florida (USBR 2006).

Regulations Governing Deep-Well Injection

Deep-well injection is regulated by the underground injection control (UIC) standards authorized under the SDWA. The UIC regulates injection of fluids in wells or dug holes that are deeper than they are wide. The intent of the regulations is the protection of underground sources of drinking water (USDWs). A USDW is defined by the USEPA as having TDS less than 10,000 mg/L.

The USEPA established the following classification system for injection wells based on the type of waste injected:

- Class I wells—those that inject hazardous or nonhazardous industrial or municipal waste,
- Class II—those that inject fluids related to the production of oil and gas,
- Class III—those that inject fluids for extraction of minerals,
- Class IV wells—those that inject hazardous waste into a USDW (now banned), and
- Class V wells—those not included in Class I, II, III, or IV.

The USEPA classifies desalination concentrate as an industrial waste; as a result, Class I wells are designated for injection of concentrate. Injection into Class I wells must occur beneath the lowermost formation of any USDW within ¼ of a mile of the well bore. Underground injection wells must be cased and cemented to protect USDWs. UIC permits also contain requirements for operating, monitoring, and reporting. The following parameters are usually monitored:

- concentrate density, viscosity, solids, and temperature;
- concentrate volume;
- pumping rates;
- injection pressures;
- dynamic wellhead pressures;
- shut-in pressures;
- pressure decline rates;
- corrosion effects; and
- wear and tear of metal components.

Design Considerations

The first and most important design consideration for deep-well injection is site selection, and there are several criteria for a suitable site. The injection zone must be below any USDW and separated from the USDW by a hydrologically impermeable formation that prevents migration of the injected fluid into the USDW. The injection zone must meet water quality and hydrologic criteria to ensure that the injected concentrate is physically, chemically, and bacteriologically compatible with the receiving water. The receiving zone must have adequate permeability and storage volume so that injection does not increase pressure in the receiving formation. The subsurface formation cannot be in a region of the country that is subject to seismic activity and earthquakes. These site requirements preclude use of injection wells in many regions of the country. Furthermore, some states categorically prohibit injection wells.

Porosity and permeability are two important hydrogeological characteristics that determine the suitability of subsurface formations for injection of wastes. The subsurface formation must have the permeability to allow movement of the concentrate into the formation, and it must have sufficient porosity to provide the storage capacity required to hold the volume of concentrate that will be generated over the life of the plant.

Porosity is the percentage of pore volume in the formation. The pores may be connected or discrete. Effective porosity is the percentage of interconnected pore volume. To be suitable for injection, a subsurface formation must have sufficient effective porosity to store large volumes of liquid. Sandstone is an example of highly porous material, and sandstone stratum generally makes a good formation for underground injection.

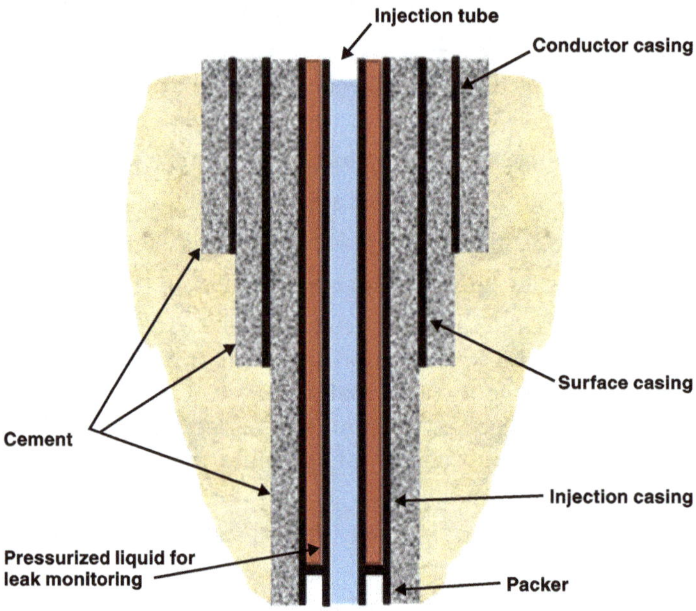

Figure 4-1 Class I injection well schematic (number of casing stages varies)

Permeability describes the ability of water to move through the formation and is related to porosity by the interconnectedness of pores in the formation. A high degree of interconnectivity facilitates water movement through the formation and increases permeability. Permeability within the formation is required, but a suitable subsurface formation must be bound by impermeable strata to prevent migration of the injected concentrate outside of the formation. A clay layer is an example of an impermeable stratum.

UIC regulations require Class I wells to be constructed with cement and casing as illustrated in the example shown in Figure 4-1. A Class I injection well comprises casing, tubing, and packers. Casing refers to an outer pipe that is cemented in place. The purpose of casing is to maintain the structural integrity of the bore hole and to seal off intermediate aquifers from the injected concentrate. Cement is used to prevent leakage, add support, and protect the pipe from corrosion. There may be several stages of casing comprising successively smaller pipe diameters with depth. Cement is placed between the outside of each casing and the sides of the bore hole and in the annular space between casings. The first and widest casing installed is the conductor casing, typically to a depth of 20–200 ft (USBR 2006). The next casing is the surface casing. The final casing is the injection casing. The injection casing contains the injection tube, the innermost pipe that conveys the injected fluid to the injection zone. Packers are used to seal the annular space between the innermost casing and the tubing, and the annulus between the casing and tubing is filled with a noncorrosive liquid maintained under positive pressure. Pressure in the annulus fluid is monitored to detect leaks. Tubing diameters are typically selected so pipe velocity does not exceed 10 ft/s.

Concentrate solutions are typically supersaturated with respect to sparingly soluble salts. Precipitation of salts after the concentrate is injected can clog pores and reduce subsurface permeability. A reduction in permeability requires higher pumping pressures to maintain the design flow rate and increases operation cost. Consequently, pretreatment with an antiscalant or acid to inhibit precipitation may be necessary prior to injection.

Cost Considerations

The cost of deep-well injection is dominated by the costs for testing and labor rather than materials (USBR 2006). As a result, larger-diameter tubing is sometimes installed to provide more capacity than is required at the time of construction. The labor cost for drilling an injection well depends on the depth of the well, tubing diameter, and soil conditions. Costs increase as well depth and tubing diameter increase. Difficult soil conditions increase labor costs.

Other factors that affect injection well cost are pumps and pretreatment requirements. Pumps are used to inject concentrate through the tubing and into the subsurface formation. Pumping pressures required for deep-well injection can vary from as little as 3 psi to as high as 5,000 psi (USBR 2006). Capital and operating costs for deep-well injection increase as pumping pressure increases.

Current Practice

Class I wells are designated for disposal of desalination concentrate, but there has been movement to allow concentrate disposal via Class II or Class V wells to make the permitting process easier and less expensive. The City of El Paso, Tex., convened a concentrate management workshop in 2012 to recommend concentrate management solutions for the city (WateReuse Research Foundation 2015). The group, comprising regulators, utility professionals, researchers, and consultants, evaluated available concentrate disposal and treatment options. The recommended option was to petition the state to create a new subclass of Class V wells designated for concentrate disposal. El Paso was granted the permit to develop Class V wells for concentrate disposal, and the Kay Bailey Hutchison Desalination Plant has been in operation since May 2012.

The USBR sponsored a study to evaluate use of Class II wells for concentrate disposal in Texas (USBR 2005). The authors noted that Class I permits require years and millions of dollars to obtain compared to months and thousands of dollars to obtain a Class II permit. The authors concluded that there were more than 31,000 oil and gas field wells in Texas, many fitting the criteria for separation from USDWs, and that using the wells would pose no risk to groundwater supplies in the state. At the time of this printing, however, there have been no Class II permits issued for concentrate disposal.

Advantages and Disadvantages

Table 4-3 summarizes the advantages and disadvantages of deep-well injection. Deep-well injection has a history of successful practice in Florida, where subsurface conditions are

Table 4-3 Advantages and disadvantages of deep-well injection

Advantages	Disadvantages
• Effective solution for regions such as Florida with ideal hydrogeological subsurface conditions • Under the right hydrogeological conditions, can provide a solution for large concentrate flows	• Lengthy and expensive permitting under current regulations requiring disposal in a Class I well • Not a viable solution where prohibited by the state, subsurface formation does not have the required permeability and porosity, or the region is subject to earthquakes • Extensive study and testing required to determine whether a subsurface formation is appropriate for deep-well injection • High operating and maintenance costs relative to other options • Potential for the permeability of the subsurface formation to be reduced by precipitation of salts in the concentrate, resulting in higher pumping pressures and costs

ideal for accepting concentrate. Under the right subsurface conditions, disposal of large concentrate flows by deep-well injection is more practical than discharge to sewer or evaporation ponds.

One reason deep-well injection of concentrate has been practiced predominantly only in Florida is that there are many factors that can preclude its use. The practice is banned in some states, and deep-well injection is not allowed in areas that are subject to earthquake. The stringent requirements for subsurface permeability, porosity, storage capacity, confinement, and separation from USDWs preclude use of many subsurface formations for deep-well injection. The permitting process for Class I wells is rigorous and can be expensive. In most cases, extensive hydrogeological study and testing is required to determine whether a subsurface formation is suitable for deep-well injection. These tests and studies can be expensive, and there is no guarantee that the formation will be deemed to be suitable. Even after pretreatment, precipitation of salts from the concentrate can occur after injection into the subsurface. Such precipitation can reduce the permeability of the formation and require greater pumping pressure and energy cost to maintain the same injection flow rate.

EVAPORATION PONDS

Discharge of concentrate to an evaporation pond can be a cost-effective concentrate management strategy for small concentrate flows in arid climates. Evaporation ponds are generally easy to construct, and operational costs are minimal because no mechanical equipment is required. However, capital costs are high relative to other discharge options and even compared to treatment options in some cases. Most states require a leak-proof liner and monitoring system to reduce the risk of groundwater contamination; these features add substantially to the construction cost of evaporation ponds. Evaporation requires large land areas because water has a high heat of vaporization. Large pond surface areas are required for effective evaporation of small concentrate flows even in arid climates.

Evaporation ponds have been used for decades to manage liquid waste from desalination, industrial treatment, and power plant facilities (Salzman et al. 2001, Ahmed et al. 2000, Westbrook and Wirth 1977), and they have been used for centuries to recover salt from seawater. The share of desalting plants using evaporation ponds appears to be on the decline as the number and size of desalination facilities increase. Mickley (USBR 2006) reported the following regarding concentrate disposal to evaporation ponds in the United States:

- In 1992, six percent of desalting plants surveyed discharged concentrate to evaporation ponds.
- By 2002, the percentage of desalting plants discharging concentrate to evaporation ponds had declined to two percent.
- Disposal to evaporation ponds was typically used by plants with less than 1 mgd capacity.

Evaporation is the process by which water is transformed from the liquid phase to the vapor phase by diffusion. Mass transfer by diffusion is driven by a concentration gradient; for evaporation, the driving concentration gradient is the difference between the actual vapor pressure of air at the air–water interface and its vapor pressure at saturation.

Saturation vapor pressure is an indication of the capacity of air to hold moisture, and it increases as air temperature increases. Hence, evaporation ponds are most effective in dry, warm climates. Even in arid regions, however, evaporation requires large surface areas. Accordingly, practical use of evaporation ponds for concentrate disposal is limited to cases with small concentrate flow rates and locations with an arid climate and relatively inexpensive land.

As water evaporates from the pond, salts in the concentrate are left behind as solids. These residual salts are either removed from evaporation ponds periodically and transported to a landfill or they are left in place and landfilled in situ by covering the pond when its storage capacity is exhausted.

Regulations Governing Evaporation Ponds

Under USEPA regulations, all waste from municipal treatment plants, including concentrate, is classified as industrial waste. There are no federal regulations governing discharge of concentrate to evaporation ponds; rather, the practice is regulated by state and local agencies. A state permit to discharge to an evaporation pond may contain effluent sampling and pond monitoring requirements for parameters such as discharge flow rate, pH, temperature, hardness, alkalinity, nitrate, sulfate, TDS, priority pollutant metals, and pond sludge volume. State agencies may also consider the impact of an evaporation pond on wildlife. One example is the deactivation of evaporation ponds in the San Joaquin Valley of California (The San Joaquin Valley Drainage Implementation Program and The University of California Salinity/Drainage Program Evaporation Ponds Technical Committee 1999). In 1985, there were 28 evaporation ponds operating in the San Joaquin Valley. By 1999, only 10 of the 28 ponds remained in operation. The others were deactivated because of the high cost of meeting effluent discharge requirements or because of toxic effects of selenium on water birds.

Design Considerations for Evaporation Ponds

Design of evaporation ponds requires determination of pond surface area and pond depth. Pond surface area requirements are based on the local evaporation rate, which in turn depends on site-specific climatological conditions including temperature, solar radiation, wind speed, and humidity (Burman and Pochop 1994). Evaporation pans are often used to simulate pond evaporation to develop site-specific design evaporation rates for pond sizing. The standard evaporation pan in the United States is the Class A pan of the National Weather Service. The Class A pan is a cylindrical container with a 4-ft diameter and 10-in. depth. Water is maintained between 2 and 3 in. from the rim, and water level, wind speed, and water temperature are measured continuously.

Because of differences in scale and sidewall effects, evaporation from a pond differs from that from a pan; a pan coefficient, defined as the ratio of pond evaporation to pan evaporation, is frequently used to correlate the two. Kohler (1954) suggested a pan coefficient of 0.7. Linacre (1994) surveyed pan coefficients reported in 18 papers and found a wide scatter around an average value of 0.77. This factor is referred to as the Class A pan coefficient.

The effect of salinity must also be accounted for in sizing evaporation ponds for concentrate management. Water vapor pressure decreases as salinity increases, causing the rate of water evaporation to decrease also. At saturation concentration for sodium chloride (26.4 percent), the evaporation rate is approximately 70 percent of the rate for freshwater (USBR 2006). The USBR (2006) measured evaporation rates for a range of NaCl concentrations and reported the following percentages relative to freshwater evaporation: 98 percent for 2 percent NaCl, 97 percent for 5 percent NaCl, 93 percent for 10 percent NaCl, and 78 percent for 20 percent NaCl. Also note that an evaporation pond is expected to reach saturation during its life cycle, and in the absence of site-specific data, a 70 percent reduction in the evaporation rate is reasonable to account for salinity effects.

The following is an example of evaporation pond loading rate calculation using pond factor and salinity factor. A pan evaporation map was used to determine an annual average pan evaporation rate of 95 in./yr for the site. From historical data, the average annual precipitation rate was found to be 7.6 in./yr. The designer decided to use a salinity factor of 0.7 and a Class A pan coefficient of 0.7. A safety factor of 1.2 was used to allow for contingencies in weather or concentrate flow rates. The calculation is as follows:

$$\begin{aligned}\text{Annual average pan evaporation rate} &= 95\ \text{in./yr}\\ \text{Annual average precipitation rate} &= 7.6\ \text{in./yr}\\ \text{Net annual average evaporation rate} &= \text{Annual average pan evaporation rate} - \\ &\quad \text{Annual average precipitation rate}\end{aligned} \qquad \text{(Eq 4-1)}$$

So, for the example in the previous paragraph:

$$\begin{aligned}\text{Net annual average evaporation rate} &= 95\ \text{in./yr} - 7.6\ \text{in./yr} = 87.4\ \text{in./yr}\\ \text{Salinity factor} &= 0.7\\ \text{Class A pan coefficient} &= 0.7\end{aligned}$$

$$\begin{aligned}\text{Adjusted net annual average evaporation rate} &= 87.4\ \text{in./yr} \times 0.7 \times 0.7 = 42.8\ \text{in./yr}\\ \text{Safety factor} &= 1.2\end{aligned} \qquad \text{(Eq 4-2)}$$

$$\text{Pond loading rate} =$$

$$42.8\,\frac{\text{in.}}{\text{yr}} \times \frac{\text{ft}}{12\ \text{in.}} \times 43{,}560\,\frac{\text{ft}^2}{\text{ac}} \times 7.48\,\frac{\text{gal}}{\text{ft}^3} \times \frac{\text{yr}}{365\ \text{d}} \times \frac{\text{d}}{1440\ \text{min}} \times \frac{1}{1.2} = 1.8\ \text{gpm/ac} \qquad \text{(Eq 4-3)}$$

Pond depth must be designed to allow freeboard for wave action and precipitation of solids. The time required for evaporation is minimized when a volume of water is spread out over a very shallow depth, but evaporation ponds with very shallow depths are more prone to liner cracking and damage and do not provide enough storage volume to be practical. In a study to examine the effect of pond depth on evaporation rate, the evaporation rate decreased by only four percent as pond depth increased from 1 to 40 in. (Mickley 2001).

The freeboard required for wave action can be estimated as follows, where H_w is wave height (ft), *W* is wind velocity (mph), and *F* is fetch (ft), defined as the straight line distance wind can blow without obstruction (US Department of the Interior 1970).

$$H_W = 0.047 \times W \times \sqrt{F} \qquad \text{(Eq 4-4)}$$

The maximum storage requirement for accumulation of solids can be estimated as follows, where *d* is the solids accumulation rate (ft/yr), *q* is the pond loading rate (gpm/ac), TDS is the concentration of total dissolved solids in the concentrate (mg/L), and ρ is the density of the precipitate (g/cm^3):

$$\begin{aligned} d\ \text{ft/yr} &= q\,\frac{\text{gal}}{\text{min}\ ac} \times 3.785\,\frac{\text{L}}{\text{gal}} \times \text{TDS}\,\frac{\text{mg}}{\text{L}} \times 10^3\,\frac{\text{g}}{\text{mg}} \div 43{,}560\,\frac{\text{ft}^2}{\text{ac}}\\ &\quad \times 525{,}600\,\frac{\text{min}}{\text{yr}} \div \rho\,\frac{\text{g}}{\text{cm}^3} \div 30.48^3\,\frac{\text{cm}^3}{\text{ft}^3}\end{aligned} \qquad \text{(Eq 4-5)}$$

Typical densities for precipitates from concentrate range from 2.3 to 2.7 g/cm^3. The depth of storage required for solids can be estimated by multiplying the solids accumulation rate, *d*, by the life of the pond in years.

Orientation of the pond can help with the evaporation rate and efficient use of freeboard. As wind blows across the water surface, the air becomes increasingly saturated with water vapor, and the rate of evaporation decreases. Therefore, the evaporation rate can be maximized by orienting the long dimension of the pond perpendicular to the prevailing wind direction. This orientation will also minimize wave height in the pond.

Operation and Maintenance

Operation and maintenance requirements for evaporation ponds are minimal compared to most concentrate disposal options. Solids that accumulate in the bottom of the pond must be cleaned out if there is insufficient pond volume for their storage over the life of the pond. The pond liner and dykes must be inspected regularly for damage. Periodic inspection of the pipe and flow control devices is also required. Monitoring wells are needed to ensure that the groundwater is not impacted by salt levels in the evaporation pond.

Cost Considerations

Evaporation pond cost includes the following:

- land cost,
- earthwork,
- pumping and piping to the pond,
- pond liner,
- leak detection and monitoring systems,
- solids disposal, and
- maintenance.

The cost of land is typically small relative to the cost of construction. Nevertheless, evaporation ponds are easier to site in regions that have large tracts of open, inexpensive land. Construction cost is affected by the topography of the land and the amount of clearing that must be performed. Earthwork is less expensive for level land. Requirements for the pond liner vary from state to state, but the pond liner is typically a significant part of the construction cost. Maintenance costs for evaporation ponds are minimal. There is a cost associated with removal of solids from the pond and their landfill disposal.

Advantages and Disadvantages

Table 4-4 summarizes the advantages and disadvantages associated with discharge of concentrate to evaporation ponds. The primary advantage of evaporation ponds is simplicity. No NPDES permit is required, and the ponds require minimal operation and maintenance. The disadvantages are the large land area required, the high construction cost, and the potential for leakage and contamination of underlying groundwater.

Table 4-4 Advantages and disadvantages of discharge of concentrate to evaporation ponds

Advantages	Disadvantages
• No NPDES permit is required. • Minimal maintenance is required. • The ponds are operationally simple.	• Large land areas are required. Consequently, this option is only practical for small concentrate flows. • Pond construction is expensive, and there is little economy of scale. • There is potential for leakage and contamination of groundwater.

ENHANCED EVAPORATION SYSTEMS

Enhanced evaporation systems use mechanical or solar energy to increase evaporation rates relative to conventional evaporation ponds. The enhanced evaporation systems that have been tested can be categorized into three general modes of operation:

- spray systems that increase the surface area available for evaporation by projecting concentrate into the air as droplets or mist,
- a film evaporation system that increases surface area available for evaporation by cascading concentrate down strips of material in a film, and
- a solar pond system that increases the rate of evaporation by using solar energy with a salt density gradient.

Spray Systems

The USBR and the Salton Sea Authority for the Salton Sea Restoration Project evaluated spray-enhanced evaporation systems of three manufacturers (2007). Each system uses mechanical energy to disperse concentrate as droplets into the atmosphere. A portion of the water is evaporated, and unevaporated water and residual salt particles fall to the ground. The percentage of water evaporated depends on the size of the droplet, air temperature, relative humidity, wind speed, and the amount of time the droplet spends in the air.

In one system, several water lines with spray nozzles (shower lines) are suspended between towers that are typically 30 m in height. Concentrate is pumped from a holding pond to the shower lines where it is dispersed to the atmosphere as a mist. A portion of the water is evaporated. Unevaporated water and residual salts fall to the ground.

Another system is a ground-based unit that uses an axial flow turbine to produce a high-velocity wind stream. Water is sprayed into the wind stream through nozzles, and water droplets are projected into the air. As with the first system, a portion of the water is evaporated and remaining water and residual salts fall to the ground.

The final spray-enhanced evaporation system that was evaluated is also ground based. The principle is similar to that of the previously described system, but in this one, water is sprayed into the air by a high-speed fan.

Each of these systems has the potential to achieve efficient evaporation in arid climates. The disadvantage of these systems is the potential for contamination of the surrounding area with salt. Salt particles can be carried long distances by wind, and there is little control of where the salt ends up. Consequently, salt drift can cause contamination of a large area surrounding the system. In the tests at Salton Sea, salt drift distances of up to 1,300 ft were measured during wind velocities of 10 mph or less.

Table 4-5 summarizes the advantages and disadvantages associated with spray systems. The primary advantage of this approach is that spray systems are relatively inexpensive to build and operate. The primary disadvantage is the salt in concentrate is not controlled, creating a high risk of contamination of the surrounding area with salt. Equipment can also become scaled with salt, causing corrosion and loss of hydraulic capacity.

Table 4-5 Advantages and disadvantages of spray systems

Advantages	Disadvantages
• Inexpensive to build and operate • Potential to achieve efficient evaporation in arid climates	• High risk of contaminating the surrounding area with salt • Maintenance required for scaling by salt

Table 4-6 Advantages and disadvantages of WAIV

Advantages	Disadvantages
• Reported to require 10 percent of the land area required by evaporation ponds • Minimal equipment and equipment maintenance • Operationally simple	• Emerging technology with no record of long-term performance or cost • Large land area required relative to several of the other discharge options • Effective only in arid climates • Management of residual salts required

Wind-aided Intensified Evaporation

Wind-aided intensified evaporation (WAIV; Gilron et al. 2003) is an enhanced evaporation system that uses vertically mounted and continuously wetted evaporation surfaces to increase surface area and enhance evaporation rates. Concentrate is pumped to the top of the strips and allowed to fall as a film down the length of the strips; thus, a high surface area per unit volume of water is exposed to the air. Water that is not evaporated falls into a holding pond under the WAIV strips and is recirculated. Researchers have reported evaporation rates 10 times that of an evaporation pond with an equivalent footprint (Katzir et al. 2009, Gilron et al. 2003). Salts that precipitate on the strips fall into the holding pond. Salts that fall from the strips or precipitate in the pond must be cleaned out periodically.

The keys to the design of WAIV are the strip material and the spacing of the strips. The material must be hydrophilic enough to allow spreading of the concentrate over the strip area but not so hydrophilic as to reduce vapor pressure. The strips must be spaced to allow sufficient air flow between strips while minimizing the footprint required. The WAIV technology is owned and sold by a private company. WAIV has been pilot tested but currently there are no full-scale installations.

Table 4-6 summarizes the advantages and disadvantages of WAIV. WAIV uses a natural renewable energy source (wind) to evaporate the concentrate using reportedly only 10 percent of the land area required by evaporation ponds. The only mechanical equipment required is a pump to move the concentrate from the holding pond to the top of the strips. Consequently, mechanical equipment operation and maintenance is minimal. Because WAIV is an emerging technology that has been used only in pilot tests, there is no record of long-term operation from which to assess treatment performance and cost. WAIV is more effective in arid climates. While it requires less space than evaporation ponds, land requirements are still considerable. As with evaporation ponds, the salts removed with WAIV must be managed.

Salt Gradient Solar Pond

A salt gradient solar pond (SGSP) uses a salinity gradient to collect and store heat that can be used for industrial and agricultural purposes (Lu et al. 2000). The University of Texas at El Paso (UTEP) operated an SGSP adjacent to a canning plant in El Paso from 1987 to 1992 (Lu et al. 2000). During this demonstration project, the SGSP was used to preheat water for a boiler, to desalinate water, and to generate electricity (USBR 2002).

Figure 4-2 shows a schematic of an SGSP. An SGSP is designed to allow it to capture and hold solar energy. The density of a saline solution increases as salinity increases, and decreases as temperature increases. The effect of salinity on density creates three vertical zones on the SGSP as higher-salinity and higher-density water collects at the bottom of the pond.

Water temperature and salinity increase with depth through the three zones. Heat is lost to the atmosphere in the upper convection zone. This zone is cooler and has the

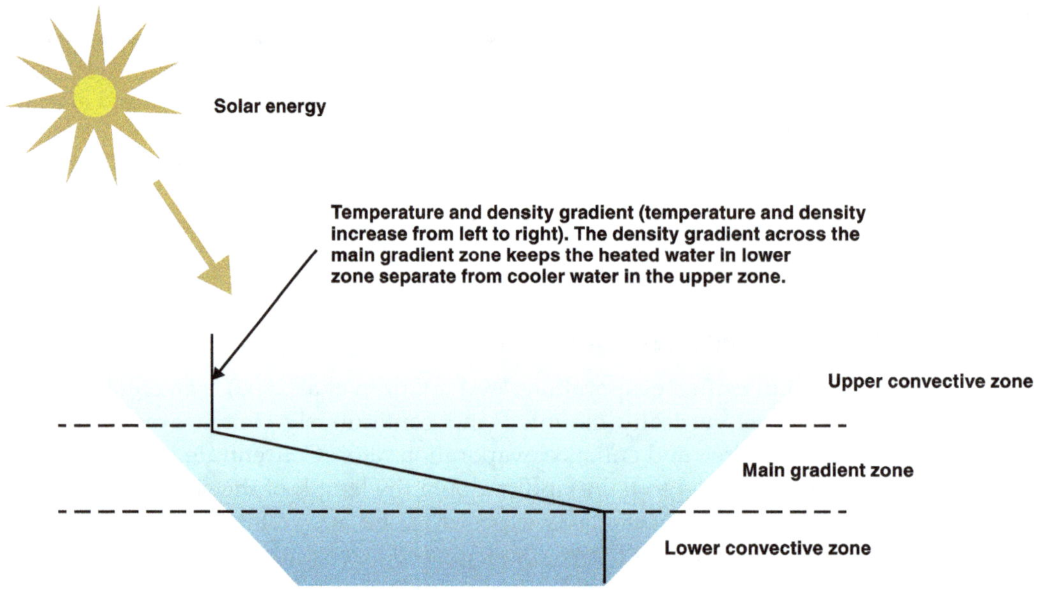

Figure 4-2 Salt gradient solar pond schematic

lowest salinity and lowest water density. The middle zone is the gradient zone. Water temperature, salinity, and density increase with depth through this zone. The bottom zone is the heat storage zone. Temperatures in the bottom zone reach 70°C–100°C. This zone has the highest water temperature, salinity, and density.

Without the vertical density gradient, heated water in the bottom zone would become lighter than the cooler water above it, causing the pond to invert, with cooler water falling to the bottom and warmer water rising to the surface, where heat would be lost to the atmosphere. In the SGSP, however, water at the bottom of the pond is held in place by the salinity-induced density gradient, and heat is collected and stored.

The UTEP researchers reported the following regarding pond performance during testing (Lu et al. 2000; Swift and Golding 1992): The pond was slightly deeper than 3 m, and it took four days to establish the salinity gradient. During measurements made on a day in September, the upper convection zone was about 0.5 m, and temperature and density were uniform through the zone at 26°C and a specific gravity of 1.01. The main gradient zone was approximately 1 m thick. Temperature increased from 26°C at the top of the zone to 86°C at the bottom of the zone. Specific gravity increased from 1.01 at the top of the zone to 1.21 at the bottom of the zone. The lower zone was approximately 1.5 m thick. Temperature and density were uniform through the lower zone at 86°C and a specific gravity of 1.21.

The average solar radiation in El Paso during testing was reported to be 5.7 kWh/m^2/d. The rate of heating the storage zone was proportional to the incoming solar radiation and inversely proportional to the thickness of the bottom layer. The thickness of the storage zone was controlled by the rate of withdrawing heat from the pond. The researchers reported that heat from the pond can be exchanged for beneficial use in two ways: (1) by pumping the heated concentrate from the bottom of the pond to a heat exchanger near the pond or (2) by pumping water through a heat exchanger located at the bottom of the pond. During the five years of operation, the thickness of the lower zone ranged from 0.75 to 1.8 m, and the temperature difference between the upper and lower zones ranged from 55 to 75°C.

The UTEP researchers coupled the SGSP with a brine concentrator and recovery system (BCRS; USBR 2002). The SGSP provided feedwater to the BCRS, and the BCRS was

Table 4-7 Advantages and disadvantages of salt-gradient solar ponds

Advantages	Disadvantages
• The ponds require minimal equipment and equipment maintenance. • They use solar energy to generate heat that can be used for industrial or agricultural purposes.	• They are only practical in warm, arid locations. • The objective is to generate energy rather than to manage concentrate. Energy can be used to help manage concentrate, but a downstream process is required. • Careful operation is required to control the salinity gradient.

driven by thermal energy from the SGSP. Salt recovered from the BCRS was returned to the SGSP to maintain the density gradient. Approximately 26 percent of the concentrate was recovered as product water in the BCRS.

The researchers defined thermal efficiency as the ratio of incident solar heat to the heat captured from the pond (Swift and Golding 1992). They reported a thermal efficiency of approximately 9 percent during the demonstration.

The researchers reported a couple of problems operating the pond (Swift and Golding 1992). Initially, the pond was operated in a pH range of 5–6. A black precipitate thought to be the result of bacterial action and water chemistry change formed. The problem was corrected by using hypochlorous acid to maintain a pH below 4. The research team also reported incidents of instability during which they had trouble maintaining the salinity gradient between the middle and lower zones.

Table 4-7 summarizes the advantages and disadvantages of the SGSP. SGSPs require minimal equipment and equipment maintenance, and they take advantage of solar energy. However, they are only practical in warm, arid climates. Furthermore, the primary function of SGSPs is to generate heat energy for use in other processes, and therefore, they are not a stand-alone method for disposing of concentrate. Rather, SGSPs can be used to generate thermal energy used by downstream processes to treat the concentrate. SGSPs require careful operation and monitoring to control the salinity gradient.

IRRIGATION

Irrigation with concentrate is the land application of concentrate to grow crops. An NPDES permit is not required for irrigation unless the potential exists for runoff to reach a receiving water. Mickley & Associates noted that, in many cases, it is less expensive to obtain a permit than it is to prove runoff will never reach a receiving water (USBR 2006). According to USEPA guidelines (EPA/625/R-95/008), spray irrigation may be viable under the following conditions:

- The amount of TDS, chloride, sodium, boron, or salinity does not exceed a level that will cause crop damage.
- There is a requirement for irrigation in the vicinity.
- A backup disposal or storage system is available for use during periods of sustained rainfall.
- The water quality of the surficial groundwater is protected from degradation.
- Local, state, and federal regulations are met.
- A system of monitoring wells is in place.

Concentrate can have adverse effects on plant growth because of osmotic effects resulting from high TDS concentrations in soil water, toxicity of specific ions, and damage to soil structure caused by high sodium levels.

Salinity in irrigation water can have an osmotic effect, which causes plants to expend more energy on adjusting salt concentrations within their tissue to obtain the water they need from the soil. This osmotic effect leaves less energy available for growth.

If growth depression is due to excessive concentrations of specific ions rather than the osmotic effects alone, growth is inhibited by specific ion toxicity. Ions that can be toxic to plants include boron, chloride, bicarbonate, and sodium.

Although boron is vital as a trace element for plants and is supplied in fertilizer, it can become toxic at levels only slightly greater than what is required for good plant growth. Symptoms of excess boron include leaf tip and marginal burn, leaf cupping, chlorosis (yellowing leaves), anthocyanin (blue and red leaves), rosette spotting, premature leaf drop, branch dieback, and reduced growth. Among the more sensitive crops are citrus trees, which show massive leaf damage at boron levels of more than 0.3 mg/L in the irrigation water. Excess boron also reduces fruit yield and induces premature ripening on other species such as kiwi.

Chloride can cause specific injury (e.g., leaf burn, chlorosis, twig dieback) to woody plant species (e.g., stone fruits, citrus, avocados), but it is not a toxic ion for vegetable, grain forage, or fiber crops. Tolerances vary among woody species and even among varieties or rootstocks within a species. These differences usually reflect the plant's ability to exclude or retard chloride accumulation.

Bicarbonate indirectly affects iron nutrition and sodicity through its effect on soil pH. Iron availability decreases with increasing pH in part because of the precipitation of iron carbonates and reduced solubility of iron oxides.

The effect of sodium can be both direct (plant accumulation) or indirect (nutritional imbalance and impairment of soil physical conditions). Direct effects (leaf burn, chlorosis, twig dieback) can occur in avocado, citrus, and stone fruit trees. Nutritional imbalance is a consequence of insufficient concentration (<1 mmol/L) of calcium or magnesium to prevent uptake and accumulation of sodium. Consequently, as sodium levels increase, the likelihood of nutritional problems increases.

Another indirect effect of high sodium content is poor physical conditions of soil (e.g., crusts, water logging, poor permeability). Almost all crops can be adversely affected. Exchangeable sodium enhances clay swelling and dispersion (disaggregation), which decreases soil permeability to water and air. Clay swelling and dispersion depend on the levels of exchangeable sodium and salinity of the irrigation water and soil solution.

The following are important criteria in the evaluation of spray irrigation:

- hydraulic loading rate;
- land requirement;
- effect of salinity, sodicity, and carbonate on soil permeability; and
- vegetation selection.

The hydraulic loading rate (HLR) is used to determine the land area required for spray irrigation, so accurately determining the HLR is a critical step in designing a spray irrigation system. The HLR depends on the permeability of the soil and the rates of precipitation and evapotranspiration; it can be calculated using the following water balance equation, where E is evapotranspiration, P is precipitation, and T is percolation:

$$\text{HLR} = E + T - P \qquad \text{(Eq 4-6)}$$

Percolation describes the rate of water infiltration into the ground. The amount of percolation depends primarily on the permeability of the soil, but other factors such as depth to water table and the duration of the drying period between spray applications must be accounted for in the calculation. Evapotranspiration is the sum of water evaporated from the plant's surface and plant transpiration. Transpiration refers to the water that is taken into the plant through its roots and released to the atmosphere through stomata in the plant leaves. Rest periods should be built into the design calculations to allow the soil to dry and get air into the pore spaces.

Percolation, evapotranspiration, and precipitation all vary seasonally. Consequently, mass balance calculations should be performed on a monthly and annual basis to ensure adequate irrigation area throughout the year. The land requirement should include space for a buffer zone around the irrigation field, storage capacity, and possibly land for capacity expansion or contingencies. Good characteristics for an irrigation field include well-drained loamy soil; a minimum of 5 ft to the water table; and ground slopes of less than 20 percent.

Water quality characteristics may rule out spray irrigation of concentrate for many applications. Sodium can break down the soil structure and reduce the permeability of soil. When calcium and magnesium are the primary cations adsorbed by the soil, the soil tends to have a soil structure that is easily tilled and permeable. The US Department of Agriculture (USDA) developed the following equation for evaluating the effects of sodium on soil permeability, where Na, Ca, and Mg are normalities in milliequivalents per liter and SAR stands for sodium adsorption ratio:

$$\text{SAR} = \text{Na} / \left[\left(\text{Ca} + \text{Mg} \right) / 2 \right]^{1/2} \qquad \text{(Eq 4-7)}$$

Concentrate with SAR values >9 can impair soil permeability and may be toxic to the plants. High concentrations of calcium and carbonate can lead to precipitation of calcium carbonate in the soil pores, causing blockage of flow paths and reduction in soil permeability. The high salinity levels in concentrate can be toxic to plants and can damage the soil for future agricultural use because of the accumulation of salt in the soil. Accumulation of heavy metals in the soil is another concern with spray irrigation.

Irrigation water that is not taken up by the plants percolates down to the water table. Therefore, the impact of concentrate on the aquifer below the irrigation field must be considered, particularly if the water table is shallow. Drainage systems may be required to capture percolated water before it reaches the water table.

Irrigation of crops can lead to salinization of the soil. This is particularly true if concentrate is used for irrigation. The traditional approach to preventing accumulation of salts in the soil is to use occasional excessive irrigation to leach accumulated salts below the root zone. The amount of excess water that must be used is referred to as the "leaching fraction." The leaching fraction depends on the concentration of salts in the soil and the concentration of salts in the irrigation water. A commonly used equation to calculate leaching fraction is a simple dilution ratio as follows, where EC is electrical conductivity:

$$\text{Leaching fraction} - \text{EC in irrigation water/allowable EC in soil} \qquad \text{(Eq 4-8)}$$

One method for classifying the salt tolerance of crops is the relative yield of a crop under saline conditions compared to its yield under normal conditions (USDA 1954). Crop yield is affected by the salinity of both the irrigation water and the soil.

A common classification for salt tolerance of plants is shown in Figure 4-3. Relative crop yield is plotted against electrical conductivity of the irrigation water. Salt tolerance is classified as sensitive, moderately sensitive, moderately tolerant, and tolerant based on the reduction in crop yield caused by irrigation water electrical conductivity. The salt tolerance of a crop is also affected by the permeability of the soil. A crop can tolerate higher-salinity irrigation water in well-drained soil better than in poorly drained soil.

Table 4-8 lists some examples of crops classified by this method. There are many more crops in each category than are listed in the table, but these examples serve to show the

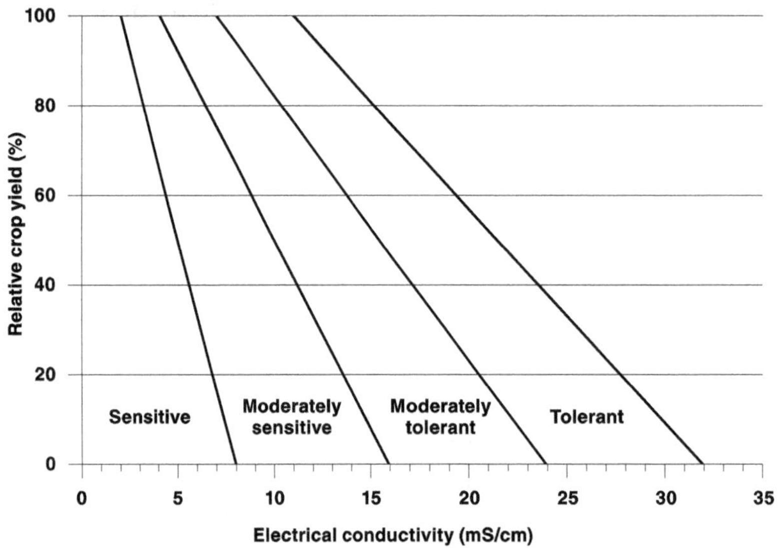

Source: Adapted from Maas and Hoffman 1977

Figure 4-3 Plant salt tolerance classifications

Table 4-8 Examples of crop salt tolerance

Classification	Examples
Sensitive	Rice, sesame, beans, carrot, onion, almond, apricot, avocado, cherry, blackberry
Moderately sensitive	Squash, sweet potatoes, tomatoes, spinach, turnips, watermelon, grapes, papaya, pecans
Moderately tolerant	Safflower, soybean, sunflower, wheat, clover, fescue, ryegrass, lima bean, beet, zucchini, turnip
Tolerant	Barley, cotton, millet, oats, rye, bermuda grass, saltgrass, wheatgrass, asparagus

Table 4-9 Advantages and disadvantages of irrigation with concentrate

Advantages	Disadvantages
• Beneficial use of concentrate to produce crops • Could replace the use of better-quality water for irrigation • Relatively inexpensive	• Potential for contamination of groundwater by infiltration and surface water by runoff • Potential to contaminate the soil with salt • Requires storage when crops cannot be irrigated • Individual substances in concentrate such as boron are potentially toxic to plants • Individual substances in concentrate may pose potential health risks to humans

potential for irrigation with concentrate if the environmental risks to the soil and adjacent water supplies can be controlled.

Table 4-9 lists the advantages and disadvantages of irrigation with concentrate. More than 70 percent of freshwater use worldwide is for irrigation, and irrigated lands supply 60 percent of the world food supply (Garces-Restrepo et al. 2007). Reliance on irrigation will increase as the world population increases. Concentrate potentially could be used to produce crops while reducing the stress on freshwater supplies; it is a relatively inexpensive option. However, there are several potential challenges that would have to be surmounted to make irrigation with concentrate practical and sustainable. The primary challenge is controlling the potential for salt contamination of the soil and surrounding water resources. Also, irrigation can be applied only intermittently. Consequently, there must be adequate land and emergency concentrate storage volume to manage the concentrate flow. In addition, there may be contaminants in the concentrate that are a concern for the health of the plants or humans when the plants are consumed.

REFERENCES

Ahmed, M.; Shayya, W.H.; Hoey, D.; Mahendran, A.; Morris, R.; and Al-Handlaly, J. 2000. Use of Evaporation Ponds for Brine Disposal in Desalination Plants. *Desalination*, 130:155–168.

Burman, R. and Pochop, L.O. 1994. *Evaporation, Evapotranspiration, and Climatic Data* (*Developments in Atmospheric Science*). Amsterdam: Elsevier.

Garces-Restrepo, C.; Vermillion, D.; and Munoz, G. 2007. *Irrigation Management Transfer*. FAO Water Reports 32. Rome: Food and Agriculture Organization of the United Nations.

Gilron, J.; Folkman, Y.; Savliev, R.; Waisman, M.; and Kedem, O. 2003. WAIV—Wind Aided Intensified Evaporation for Reduction of Desalination Brine Volume. *Desalination*, 158(1–3):205–214.

Katzir, L.; Volkmann, Y.; Daltrophe, N.; Korngold, E.; Mesalem, R.; Oren, Y.; and Gilron, J. 2009. WAIV—Wind Aided Intensified Evaporation for Brine Volume Reduction and Generating Mineral Byproducts. *Desalination and Water Treatment*, 13(1–3):63–73.

Kohler, M.A. 1954. *Lake and Pan Evaporation*. pp. 127–148. https://pubs.usgs.gov/pp/0269/report.pdf (accessed Dec. 3, 2018).

Linacre, E.T. 1994. Estimating US Class A Pan Evaporation From Few Climate Data. *Water International*, 19(1):5–14.

Lu, H.; Walton, J.C.; and Swift, A.H.P. 2000. Desalination Coupled With Salinity-Gradient Solar Ponds. *Desalination*, 136(1–3):13–23.

Maas, E.V. and Hoffman, G.J. 1977. Crop Salt Tolerance: Current Assessment. *Journal of the Irrigation and Drainage Division* (*ASCE*), 103:115–134.

Mickley, M. 2001. Membrane Concentrate Disposal: Practices and Regulations. Desalination and Water Purification Research and Development Program Report No. 69. Washington: USBR.

Robert, C. 2010. Reverse Osmosis Design & Concentrate Discharge Evolution in Florida the Past Three Decades. *Florida Water Resources Journal*, November 2010.

Salzman, S.A.; Allinson, G.; Stagnitti, F.; Coates, M.; and Hill, J.R. 2001. Performance of Constructed Evaporation Ponds for Disposal of Smelter Waste Water: A Case Study at Portland Aluminum, Victoria, Australia. *Water Research*, 35(9):2121–2128.

Swift, A.P. and Golding, P. 1992. Topics in Gradient Maintenance and Salt Recycling in an Operational Solar Pond. *Journal of Solar Energy Engineering*, 114(1):62–69.

The San Joaquin Valley Drainage Implementation Program and The University of California Salinity/Drainage Program Evaporation Ponds Technical Committee. 1999. *Evaporation Ponds*. https://water.ca.gov/LegacyFiles/pubs/groundwater/evaporation_ponds_final_report__san_joaquin_valley_drainage_implementation_program/06-evapponds.pdf (accessed Feb. 21, 2019).

USBR (US Bureau of Reclamation). 2006. Membrane Concentrate Disposal: Practices and Regulation. Washington: USBR.

USBR. 2005. Please Pass the Salt: Using Oil Fields for the Disposal of Concentrate from Desalination Plants. Desalination and Water Purification Research and Development Program Report No. 112. Washington: USBR.

USBR. 2002. Zero Discharge Waster Brine Management for Desalination Plants. Desalination and Water Purification Research and Development Program Report No. 89. Washington: USBR.

USBR and Salton Sea Authority. 2007. *Restoration of the Salton Sea*. https://www2.usgs.gov/saltonsea/docs/reclamation/Reclamation%20FinalSummaryRpt.pdf (accessed Dec. 3, 2018).

USDA (US Department of Agriculture). 1954. *Agriculture Handbook No. 60: Diagnosis and Improvement of Saline and Alkali Soils*. Washington: USDA.

US Department of the Interior. 1970. Disposal of Brine by Solar Evaporation: Field Experiments. Washington: US Department of the Interior.

USEPA (US Environmental Protection Agency). 2010. *NPDES Permit Writers' Manual*. EPA-833-K-10-001. Washington: USEPA.

USEPA. 1991. *Technical Support Document for Water Quality–based Toxics Control*. EPA/505/2-90-001. Washington: USEPA.

WateReuse Research Foundation. 2015. Framework for Direct Potable Reuse. Project 14-20. Alexandria, Va.: WateReuse Research Foundation.

Westbrook, G. and Wirth, L. Jr. 1977. Water Management at Power Plants. *Environmental Science & Technology*, 11(2):140–143.

Chapter 5

Enhanced Recovery and Zero Liquid Discharge

Brent Alspach and Graham Juby

Because desalination is among the most expensive treatment processes, it is typically one of the last alternatives for generating potable water supplies. Consequently, for those cases in which desalination is used, maximizing the recovery of the process is often an important consideration. In addition to enhancing yield, maximizing recovery reduces the desalination concentrate flow, which may improve the diversity, feasibility, and/or economics of management options. This chapter discusses processes that can be employed to increase recovery above what would otherwise be possible using a primary desalination step alone, including "enhanced recovery" processes and ZLD systems. Although the content of this chapter has been developed in the context of using RO technology for primary desalination, this assumption does not necessarily preclude applicability to other primary desalination processes that could be employed.

ENHANCED RECOVERY PROCESSES

As defined in this manual, "enhanced recovery" means increasing the recovery beyond that which would typically be possible for the given feedstream water chemistry, with percentages in the upper 90s. For most waters encountered in municipal treatment applications, the recovery would be between roughly 75 and 85 percent. Typically, recovery is dictated by the least soluble salt, which is often calcium carbonate, calcium sulfate, or silica. Further discussion of recovery and the factors that affect it is provided in Chapter 3.

To maximize recovery, the primary desalination process is operated just below the precipitation point of one or more of the sparingly soluble constituents. The addition of acid and antiscaling chemicals can allow operation beyond this precipitation point, increasing recovery and helping to protect the membranes from scale formation. Increasing recovery beyond that achievable in the primary desalination step in conjunction with these pretreatment chemicals requires additional treatment measures. Such enhanced recovery processes are described in this section, along with the beneficial use of any solids that may be produced.

Enhanced recovery can be considered as a progressive treatment approach, with each step contributing to increasing the recovery and providing the potential to recover more usable water; as the recovery increases, the concentrate volume decreases. However, for all such enhanced treatment trains discussed in this section, there is still at least one concentrate stream that requires management and disposal. Further concentration to the point of effectively achieving 100 percent recovery would constitute a ZLD process, which is discussed later in this chapter.

Some of the most prominent enhanced recovery processes are discussed in the following subsections, including both proprietary and nonproprietary technologies. Some strategic combinations of processes are also addressed. Note that this section is not intended to present an exhaustive overview of enhanced recovery methods, but a strategic sample of processes to provide the reader with a broad overview of applicable technologies. Additional research and development of both new and current processes and technologies is ongoing.

Chemical Softening

While chemical softening is a well-known process that was first used in water treatment to reduce hardness in England in 1841 (USBR 2010), its use to soften desalination concentrate streams is a more recent development. The purpose of chemical softening is to reduce the precipitation potential of calcium carbonate, calcium sulfate, or silica. Softening is achieved by increasing the pH of the water and providing conditions that allow precipitation to occur in a controlled manner. For example, by lowering the concentration of calcium through chemical softening, the calcium carbonate precipitation potential of a concentrate stream that is nearly saturated could be lowered significantly, which can allow increased recovery through subsequent treatment processes.

Chemical softening involves the addition of lime, caustic soda, and/or soda ash, depending on the chemistry of the water and the degree of softening to be achieved. By adding lime, either as quick lime (CaO) or calcium hydroxide [$Ca(OH)_2$], the calcium concentration of the water increases to the point of inducing calcium carbonate ($CaCO_3$) precipitation, exhausting carbonate alkalinity. The optimum pH for this reaction is about 10.3 (USBR 2010). Soda ash (Na_2CO_3) can be added to provide additional carbonate for further calcium precipitation. The degree of calcium removal can be targeted by adjusting the lime and soda ash dosing.

If removal of magnesium is also desired, then the pH needs to be increased to the range of 11.0–11.3, at which magnesium hydroxide [$Mg(OH)_2$] will precipitate. Some desalination concentrate streams contain elevated levels of silica, which can limit process recovery; however, lime-soda softening can also reduce silica concentrations at pH levels exceeding 11.

As an alternative to using lime, sodium hydroxide (i.e., caustic soda, $NaOH$) can be used to increase the pH to the point of converting bicarbonate to carbonate, thus precipitating calcium carbonate. Once the carbonate is fully reacted (i.e., precipitated), and

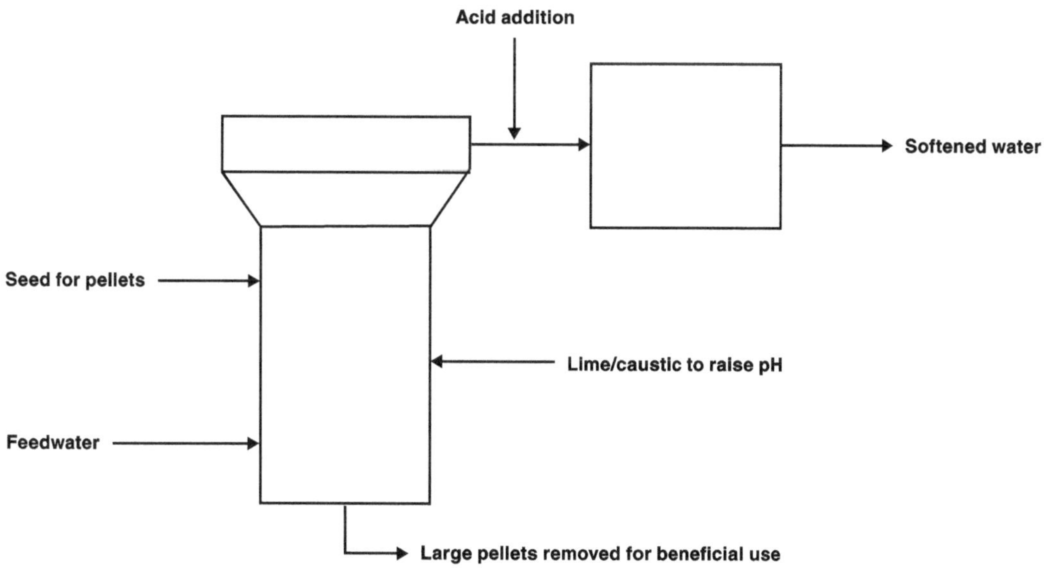

Figure 5-1 Schematic of pellet-softening process

if the pH is high enough, calcium hydroxide and magnesium hydroxide can also form. These precipitates are much more "fluffy" in nature and do not settle as easily as calcium carbonate, potentially complicating the process of removing these solids.

Another alternative to conventional lime-soda softening, albeit using the same chemistry, is "pellet softening," a fluidized reactor process in which precipitates are formed on calcium carbonate pellets that are grown on sand particulates or almost any other fine granular insoluble mineral (Faust and Aly 1998). This approach was commercialized through the Spiractor® and Crystalactor® technologies (Giesen 1999, Faust and Aly 1998). The application of pellet softening to desalination concentrate has been tested successfully at pilot scale in several locations and is currently in construction at one full-scale plant in southern California.

Figure 5-1 shows a schematic of an upflow pellet softener. In the fluidized reactor, the largest pellets migrate toward the bottom of the reactor and are intermittently removed; fresh sand or small calcium carbonate seeds are added to replace the removed pellets. Two significant advantages of pellet softening are (1) the pellets can be dewatered by gravity to 98–99 percent dryness, substantially reducing the volume of solids over that generated by conventional softening (thus reducing transportation and/or disposal costs); and (2) the pellets can be put to beneficial use, thereby enhancing process sustainability, lowering residuals management costs, and potentially generating a small revenue stream.

One of the challenges of chemical softening is the large amount of solids produced. The solids yield is particularly high in desalination concentrate applications, as the concentrations of calcium can be an order of magnitude higher than those encountered in typical hard waters that serve as raw potable sources. However, the recovered solids may potentially be recycled for beneficial use as soil conditioner, chemical additives in industrial processes, or a source of calcium and magnesium. (Salt recovery and beneficial use are discussed in more detail in Chapter 9.)

Downstream of the chemical softening process, the clarified water requires filtration and often pH adjustment prior to further treatment to enhance recovery. The extent of the filtration step depends on the softening technology used and the filterability of any remaining particulates. The combination of a seeded reactor followed by membrane filtration is a robust example of particulate removal that may allow the broadest range of

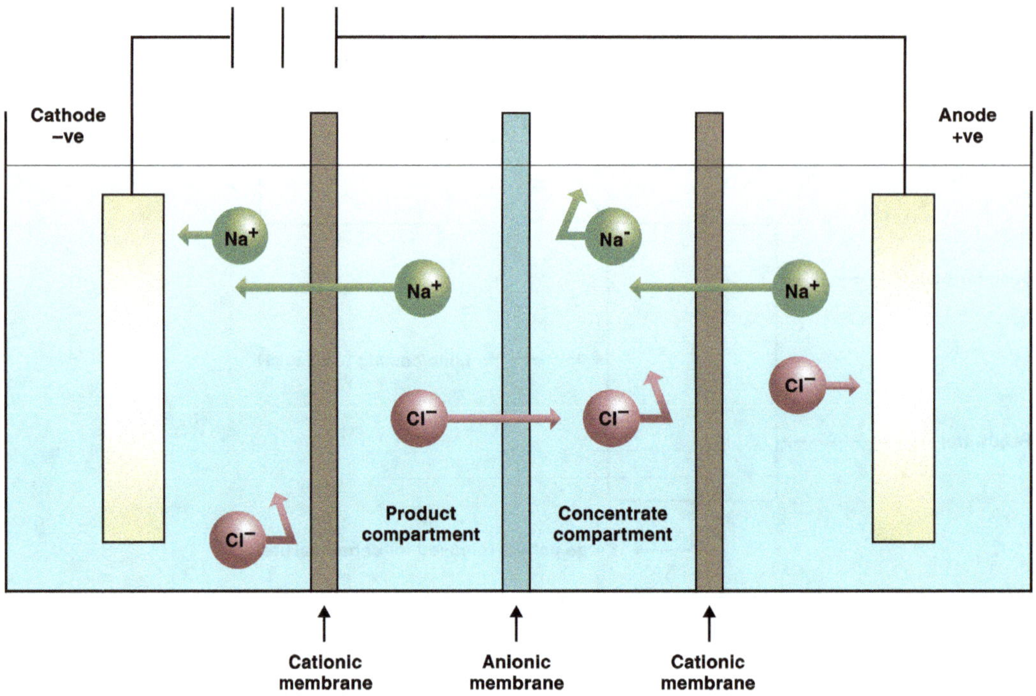

Figure 5-2 Schematic of an EDR system

upstream softening processes. Pilot testing for the chemical softening and the associated downstream filtration systems is often necessary prior to full-scale implementation to confirm process design parameters.

Secondary EDR

EDR is an electrochemical separation process that uses a DC voltage and ion-exchange membranes to desalinate water. In an EDR process, alternating cationic and anionic membrane pairs create product and concentrate compartments within a membrane stack. A schematic illustrating the principles of the EDR process is provided in Figure 5-2. In the schematic, the membranes are shown immersed in a sodium chloride (NaCl) solution. When a DC voltage is applied, the ions begin to migrate. One compartment becomes significantly depleted of ions as both positive and negative ions can pass through the cation and anion selective membranes, respectively, yielding desalinated product water. The adjacent compartment becomes more concentrated because of the influx of ions from both sides; the ion-selective membranes trap the ions in this compartment, generating the concentrate. In an EDR stack, the cell pairs are repeated many times to achieve the desired salt removal and product water recovery. An example of an EDR process flow diagram is provided in Figure 5-3. Most of the combined concentrate flow is recycled to continuously collect more ions. To control the TDS levels in the concentrate loop and avoid scale formation within the membrane stack, a portion of the flow is wasted as concentrate blowdown, and this volume is replaced with EDR feedwater. Additional detail on EDR is provided in Chapter 3 in the context of using the technology as a primary desalination process.

Depending on the chemistry of the primary desalination concentrate stream, EDR can be an effective secondary treatment step to increase recovery in a single process. For example, because noncharged species pass through the EDR stack without becoming more concentrated, feedwater at or near saturation with silica (which is uncharged at the

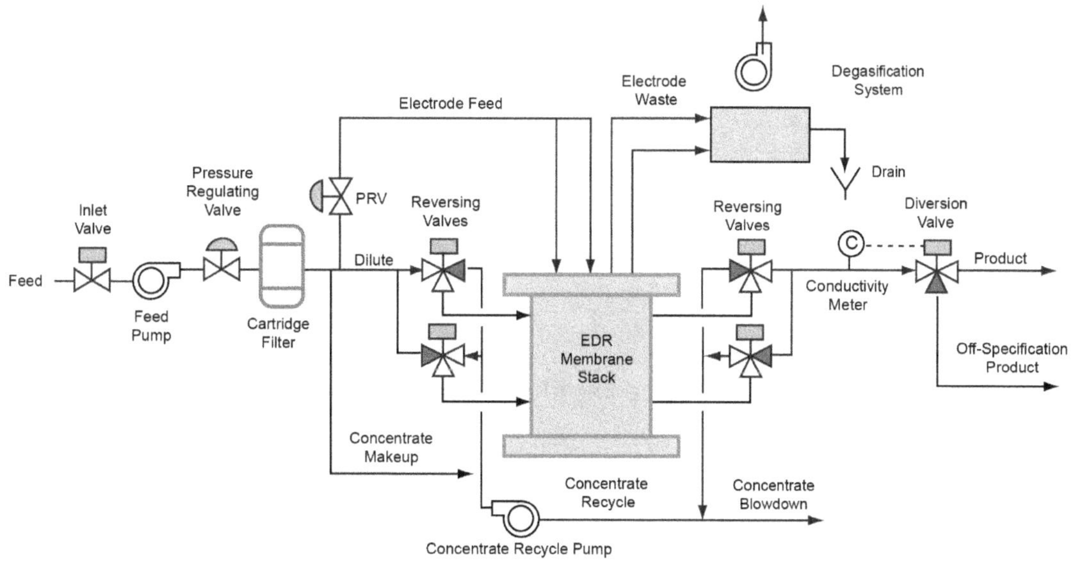

Figure 5-3 Process flow diagram of an EDR system

neutral pH range) may be well-suited for secondary treatment by EDR alone. A recent pilot test on RO concentrate from the Temescal Desalter in Corona, Calif., demonstrated that EDR achieved stable operation at a recovery of 65 percent with a silica concentration of 160 mg/L generated by the primary desalination process operating at 80 percent recovery (Wiesner et al. 2015). Combined, the two processes achieved 93 percent recovery.

More detail on EDR and related electrically driven desalination processes is provided in AWWA M38, *Electrodialysis and Electrodialysis Reversal* (AWWA 1995).

Chemical Softening Plus Secondary Reverse Osmosis or Electrodialysis Reversal

After the concentrate stream has been softened to reduce the concentration of the scale-forming constituents such as calcium and silica (as discussed previously in this chapter), additional low-salinity water can be recovered through further processing by a secondary RO or EDR process. Figure 5-4 shows a treatment schematic for this combination of processes. The "softening process" could include the use of lime/soda, pellet softening, or a similar process.

Key to the operation of these configurations is the concentration of residual dissolved solids in the feedstream to the secondary RO and EDR processes following softening. Typically, pH adjustment and solids removal steps are incorporated downstream of the softening process (as shown in Figure 5-4) to condition the water for further treatment. Lowering the pH reduces the precipitation potential of calcium carbonate, which would otherwise be at equilibrium at its precipitation potential following softening. As discussed previously, the use of sodium hydroxide can result in the formation of very fine "fluffy" precipitates that may be more challenging to remove. Because of the configuration of the membrane sheets and the larger flow area in the EDR process, EDR can typically handle a slightly higher suspended solids loading in the feedstream than can RO. Recoveries of 70–78 percent in the secondary desalting step have been demonstrated at pilot scale (Mansell et al. 2014, Juby et al. 2008).

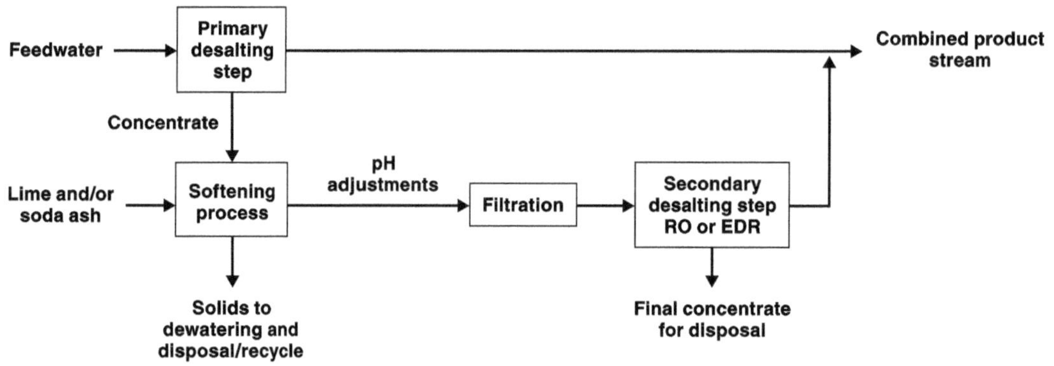

Figure 5-4 Schematic of lime/soda softening plus secondary RO or EDR

Secondary Electrodialysis Metathesis

Electrodialysis metathesis (EDM) is an electrically driven membrane desalting process similar to ED and EDR. Like ED/EDR, the EDM technology uses both cation- and anion-selective membranes to allow charged species to migrate across a barrier when a DC potential is applied. However, unlike ED/EDR, the EDM technology incorporates two feedstreams (the brackish water source water feed and a sodium chloride feed), and membranes are grouped in "quads" rather than pairs. The quad arrangement results in four product streams from the stack: desalinated water, a diluted sodium chloride stream, and two concentrate streams. As shown in the process schematic in Figure 5-5, one of the two concentrate streams becomes more highly concentrated with sulfate and sodium species, while the other becomes more concentrated with calcium, magnesium, and chloride. Separation of the calcium and sulfate ions into separate streams allows the EDM process to operate at recoveries of up to more than 90 percent. Drawbacks of the process include the additional cost of a high-concentration sodium chloride stream and management/disposal of the two highly concentrated residuals streams.

Secondary EDR Plus SPARRO

The seeded precipitation and recycle reverse osmosis (SPARRO) process was first developed to treat cooling tower blowdown from power plants that is high in calcium and sulfate ions; it has been tested in the mining industry to treat water containing high concentrations of calcium sulfate since the mid-1980s (Juby et al. 1996). Although the SPARRO process itself has not been widely used in water treatment, the concept of seeding is both well-known and well-practiced, including use in desalination via vapor compression evaporation technology (O'Neail et al. 1981).

The SPARRO process is a hybrid of conventional RO technology, incorporating the recirculation of seeded slurry through the RO system and promoting homogeneous nucleation and precipitation from the solution. Seed crystals (gypsum) are introduced to the feedstream, which is then pumped into tubular RO membranes. As the water is concentrated along the membranes, the solubility products of calcium sulfate, silicates, and other scaling salts are exceeded, and they preferentially precipitate on the seed material rather than on the membranes. Figure 5-6 shows a schematic of the SPARRO process.

The SPARRO process can be integrated with EDR to treat RO concentrate and increase recovery. The combination of EDR and SPARRO overcomes some of the limitations of both processes, increasing the recovery of the EDR unit beyond that which can be achieved by EDR alone and applying the SPARRO technology to a low-volume, high-salinity stream—a more economical approach for tubular RO systems, which have a low membrane–area–to–process–footprint ratio. As Figure 5-7 shows, feedwater from the EDR concentrate

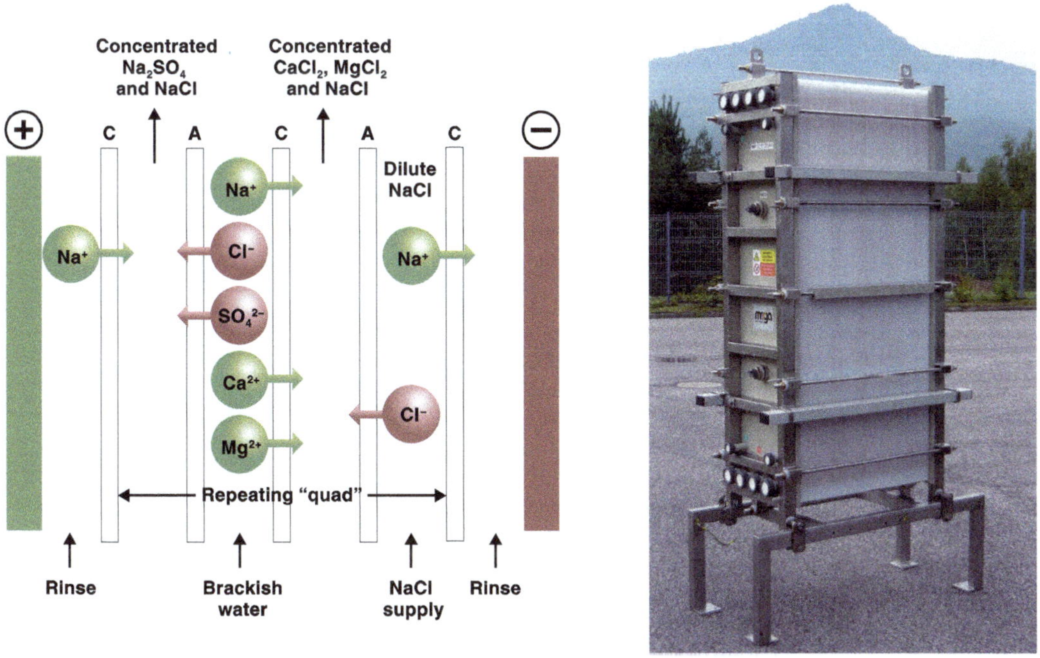

Adapted with permission from Veolia Water

Figure 5-5 EDM process schematic and picture of full-scale stack

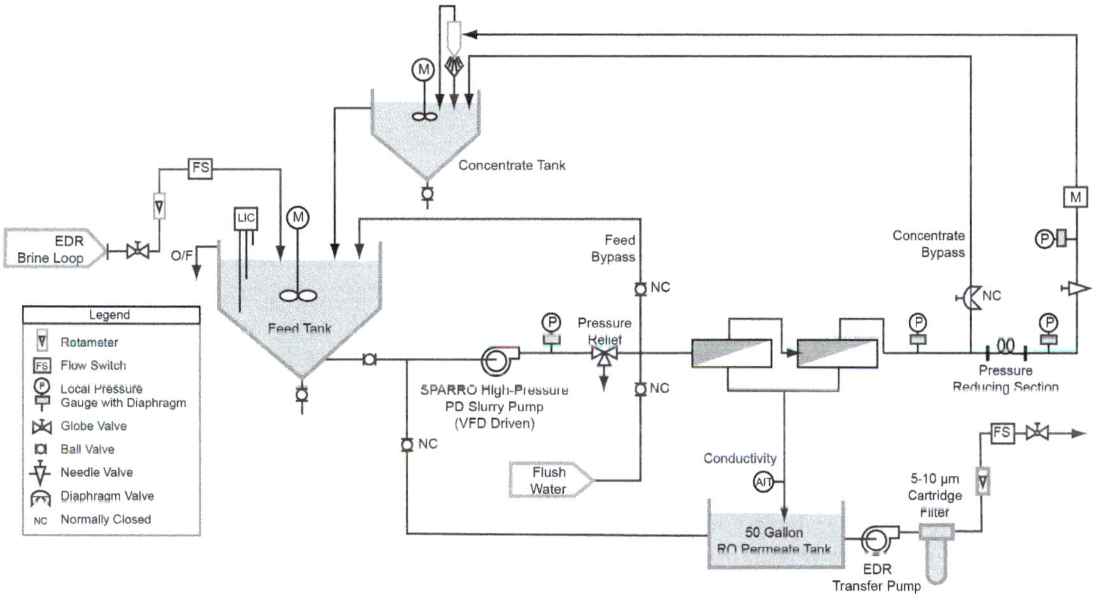

Figure 5-6 SPARRO process schematic

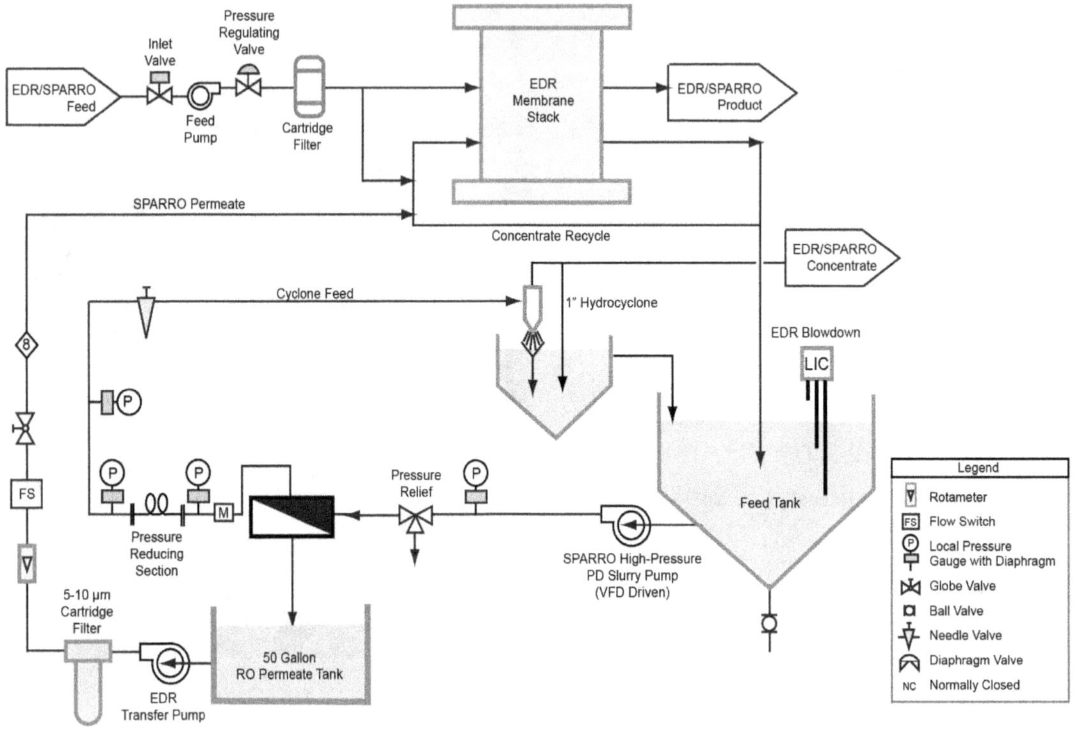

Figure 5-7 Integrated EDR-SPARRO process schematic

loop is fed to the SPARRO feed tank containing gypsum seed crystals. The slurry mixture in the feed tank is then fed to the tubular RO membrane elements. The RO permeate is collected and returned to the EDR concentrate loop, and the concentrate/slurry mixture is recycled back to the feed tank through a cyclone separator. This separation step allows for individual control of the gypsum solids mass balance and liquid TDS by wasting calculated volumes of the high-suspended-solids cyclone underflow and the high-TDS cyclone overflow, respectively.

Membrane Distillation

Membrane distillation (MD) is a process that combines simultaneous mass and heat transfer through a hydrophobic microporous membrane. Direct contact MD (DCMD) is one of four basic configurations of MD, which involves transport of water vapor through the membrane pores. Both the feed and permeate solution are kept in contact with the membrane, and its hydrophobic nature prevents the penetration of aqueous solution into the pores, resulting in a vapor–liquid interface at both pore entrances and exits. An imposed temperature difference between the two solutions gives rise to a transmembrane pressure gradient that drives the flux across the membrane. Water evaporates from the liquid–vapor interface on the feed side of the membrane and diffuses and/or convects across the membrane on the permeate side. Figure 5-8 shows a schematic of the vacuum-enhanced direct contact MD process.

Forward Osmosis

Osmosis is the naturally occurring process that biological cells use to control salinity, using osmotic pressure differences as the driving force for water transport across a semipermeable membrane to equalize solution concentrations. The mechanically facilitated RO process uses hydraulic pressure to drive water against the natural osmotic gradient to generate

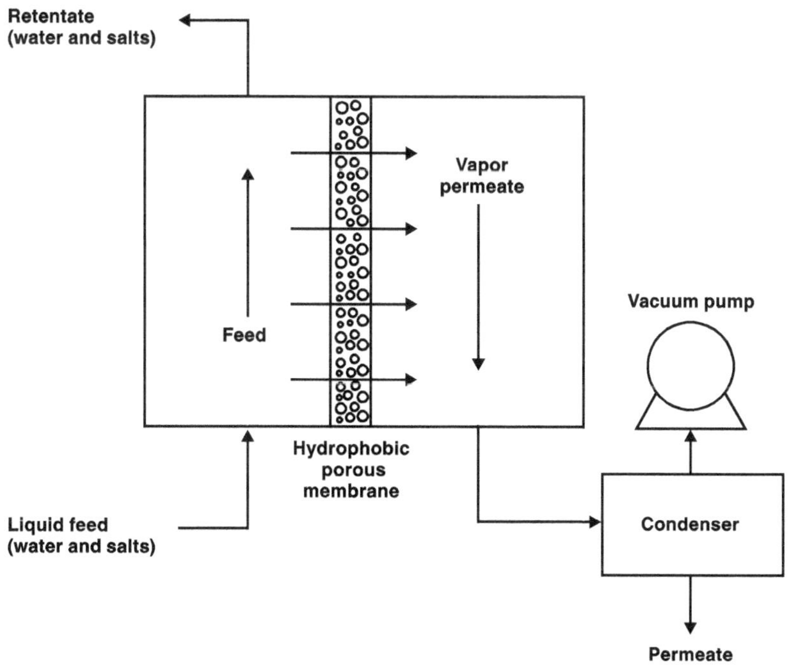

Figure 5-8 Vacuum-enhanced direct contact membrane distillation process schematic

both a low-TDS permeate stream and a highly concentrated residuals stream. The term "forward osmosis" (FO) was coined to describe an engineered process to leverage the natural osmotic gradient to accomplish desalination, inducing the flow of water from a saline feedwater to a designed hypersaline solution (i.e., a "draw solution") composed of a solute that can be more readily removed, generating freshwater. Because only standard RO membranes were available for early FO testing and concept development, fluxes were low and the process was not optimized. However, research conducted during the past decade has yielded significant advancements in FO-specific membranes, resulting in fluxes comparable to those of RO systems, and future developments may surpass these levels (Elimelech 2014). The application of FO is typically best suited for difficult-to-treat concentrate streams that cannot readily be treated by conventional processes.

Commercially, there are several ways to configure the FO technology for primary RO concentrate recovery. One recent example is the Membrane Brine Concentrator (MBC™) process that is being applied at full scale (120 gpm) to a flue-gas desulfurization concentrate stream in a power plant in China. The treatment scheme, shown schematically in Figure 5-9, includes upstream chemical softening, granular media filtration, ion exchange, and RO (not shown in the figure). The RO concentrate stream (the 75,000 mg/L feedstream in Figure 5-9) is further concentrated through an FO process that uses a proprietary high-salinity ammonium bicarbonate draw solution to create the osmotic pressure to drive water from the RO concentrate. The resulting concentrate has a TDS of around 200,000 mg/L and flows directly to a crystallizer for final disposal. The dilute draw solution from the FO process is regenerated in a thermal recovery process that yields a lower-salinity stream (~1,000 mg/L), which is recycled to the upstream side of the RO. This treatment approach replaces a thermal brine concentrator prior to the crystallizer (both discussed previously in this chapter) in achieving near ZLD.

Vibratory Shear Enhanced Process (VSEP®)

Vibratory Shear Enhanced Process (VSEP®) is a membrane separation process that employs mechanical vibration to prevent fouling. Although the concept was developed

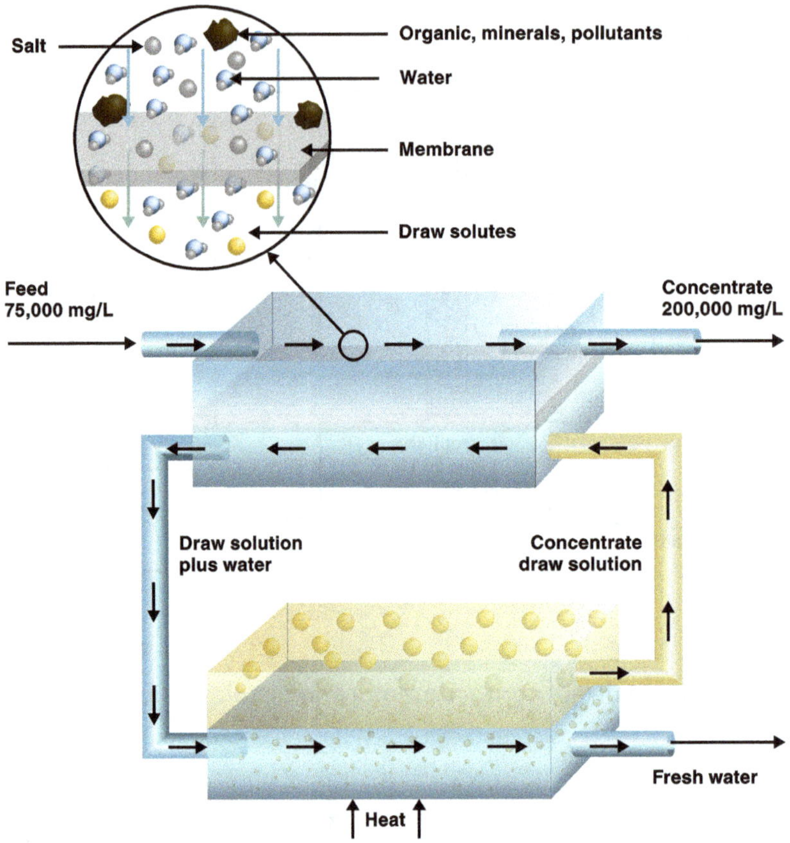

Courtesy of the Water Desalination Report, March 5, 2012
(as adapted from Oasys Water)

Figure 5-9 Membrane brine concentrator process schematic

in 1987 for use in conjunction with membrane filtration in dewatering applications, it is now applied to RO to allow the treatment of more challenging feedwater supplies. The process uses an eccentric weight bearing in communication with the membrane module or pressure vessel (see Figure 5-10), which induces a vibratory action that moves back and forth 54 times per second. The shear created by the rapid change in direction makes it difficult for foulants to attach to the membrane surface. Figure 5-11 depicts a full-scale VSEP system. Preliminary results suggest that the VSEP technology can reduce the volume of RO brine by 95 percent or more (Benjamin et al. 2010).

High Efficiency Reverse Osmosis

The High Efficiency Reverse Osmosis (HERO^{TM}) system is a proprietary technology originally developed to provide high-purity water for the microelectronics industry. The process operates at a high water recovery, typically in the range of 90–98 percent, facilitated by the specialized chemical pretreatment (GE 2010, FEMP 2004). Figure 5-12 shows a simplified schematic of the treatment process. The first step is softening to remove calcium and magnesium hardness using weak acid cation exchange resin. Acid may be added after the ion-exchange step to convert any remaining alkalinity to CO_2, which is then removed by degasification in the second step. The pH of the water is subsequently raised to 10.5 to both increase the solubility of silica and inhibit bacterial growth in the RO feed.

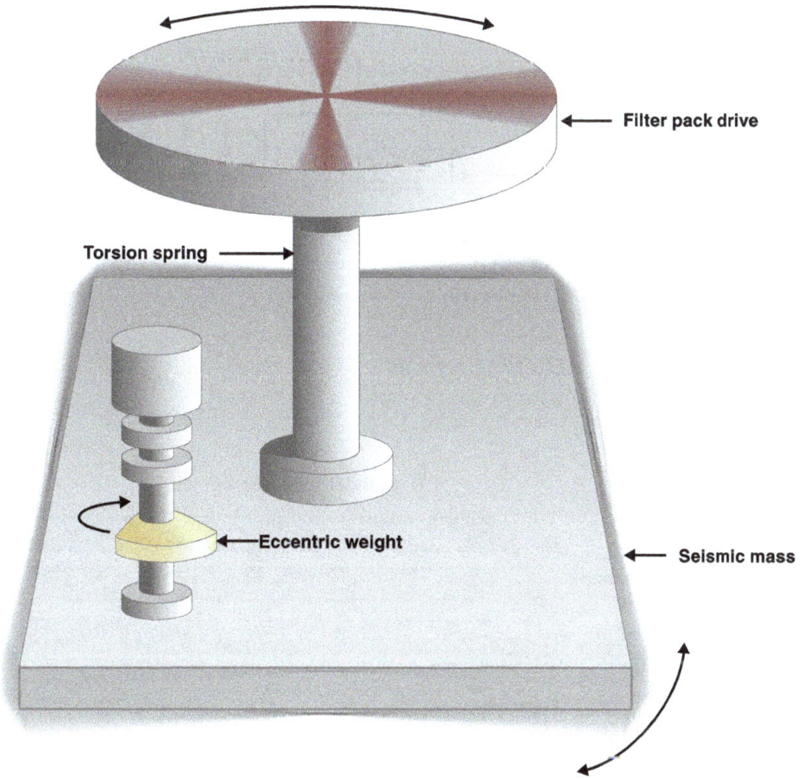

Adapted with permission from New Logic Research

Figure 5-10 Concept of the VSEP

Courtesy of New Logic Research

Figure 5-11 Example of the VSEP configuration

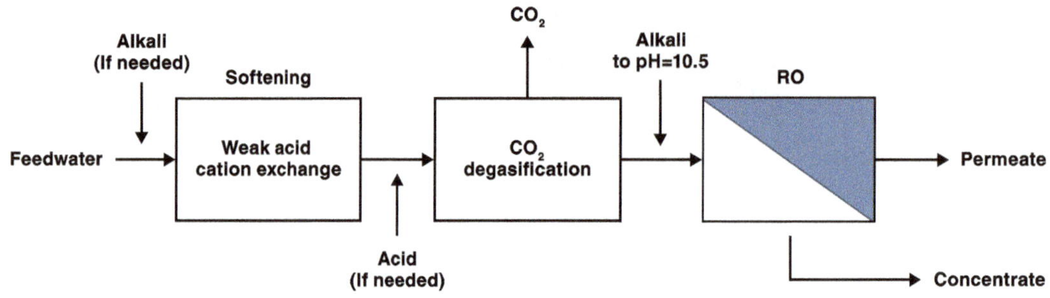

Figure 5-12 Schematic of the HERO™ process

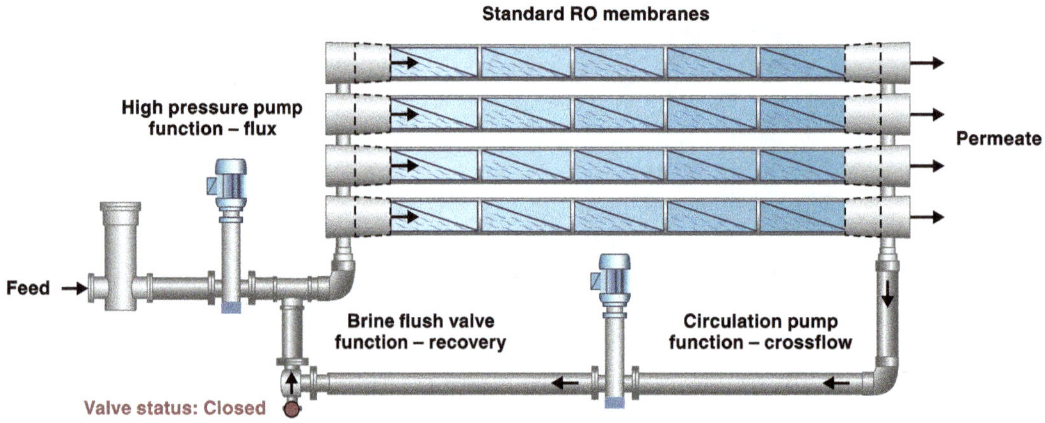

Adapted with permission from Desalitech, Inc.

Figure 5-13 Overview of the CCRO process

Closed Circuit Reverse Osmosis

A conventional RO process relies on the use of sequential membrane stages to maximize recovery in a continuous flow configuration. However, a proprietary variant called Closed Circuit RO (CCRO) instead uses a semibatch process, recycling concentrate through a single stage array until further permeate production becomes infeasible, at which time the concentrated waste is discharged. Constant variations in flow and salinity inhibit fouling and scaling, extending membrane life, and the process automatically adapts to variations in feedwater conditions, adjusting recovery goals to suit the feedwater chemistry.

Figure 5-13 shows a schematic of the CCRO configuration. The process uses the same RO membrane elements and pressure vessels as a conventional RO system, although the number of elements per pressure vessel may be lower depending on application-specific water quality and treated water objectives. As Figure 5-13 shows, concentrate from the tail end of each pressure vessel in the stage is recirculated using a separate pump and mixed with the feed downstream of the high-pressure feed pump. While recirculation is occurring, the system operates at 100 percent recovery and progressively concentrates the feedwater TDS. The recirculation process continues until a predetermined recirculation concentration is achieved, as controlled by the feedwater chemistry and monitored by a conductivity meter. At this point, the entire concentrate stream is discharged from the system by opening the brine flush valve (shown closed in Figure 5-13), and the cycle begins again. By operating in this manner, the process is controlled so that blowdown occurs only when the highest possible recovery is reached. If water quality should change, the set point on the conductivity meter can be adjusted such that blowdown will occur earlier or later (depending on the new water chemistry), with commensurate change in recovery. The CCRO configuration can operate between about 75 and 98 percent recovery, depending on the chemistry of the feedwater—a much wider operating range than is possible with conventional RO systems.

Similar to conventional two- or three-stage RO system designs, the overall recovery that can be achieved using CCRO is controlled by the feedwater chemistry and the effectiveness of the antiscalant used. However, the CCRO process has demonstrated the ability to operate at significantly higher silica concentrations than could be tolerated in conventional RO systems; because of the induction time of the silica scale, a concentrate stream supersaturated with silica can be discharged from the system before scale can form. Note that the increase in feedwater TDS over a CCRO cycle dictates not only an increase in permeate TDS, but also that the membrane elements are subject to operation under a wide range of water quality conditions.

The operation of the CCRO process also takes advantage of natural osmotic phenomena to exert an additional cleaning step that is not available to conventional RO systems. When the brine blowdown valve opens and concentrate is discharging, the system pressure drops, thus curtailing permeate flow across the membranes. Under these conditions, permeate already present on the product side of the membrane begins to flow by osmosis back across the membrane barrier, resulting in a passive flush that can help reverse and control fouling.

ZERO LIQUID DISCHARGE

The concept of ZLD represents the complete (or near complete) elimination of liquid waste from a water treatment residuals stream. For desalination processes, ZLD further concentrates the highly saline liquid residuals stream, ultimately generating solids or a dense slurry that can be transported offsite for disposal or recycling. However, the idea of ZLD is not exclusive to desalination; surface water treatment plants (i.e., those treating low-salinity sources) with advanced residuals management can likewise achieve ZLD, albeit using much different processes. Because the literature may not clearly make this distinction, and ZLD associated with low-salinity water treatment is generally much more economical and therefore more common, it is important to understand the context in which the term is applied. In accordance with the focus of this manual, the discussion of ZLD in this section is limited to applications in desalination treatment.

Because there is no formal definition of the concept, what constitutes ZLD is somewhat fluid; thus, there is likewise no widely recognized threshold recovery beyond which a desalination plant qualifies as a ZLD facility. Accordingly, the literature may represent plants operating at recoveries around 98.5 percent as using ZLD (or near-ZLD). However, accomplishing nearly 100 percent recovery is not necessarily a prerequisite for ZLD, as some water may be incorporated into hydrated solids or thickened slurries that are disposed of or recycled offsite. In such examples, the plant does not discharge any strictly liquid residuals, although technically 100 percent recovery is not attained. In any case, labeling a desalination plant or residuals management process as ZLD is less important than achieving the goal of eliminating or greatly reducing the volume of liquid residuals discharge.

There are two primary advantages associated with eliminating liquid discharge: (1) maximizing the yield of treated/desalinated water and (2) providing a means of concentrate management for cases in which there are no other viable options. The benefit of increasing treated water production could be a significant consideration in any location, potentially driven by issues of water rights, the cost of developing new sources, the availability of additional supplies, procuring land for new treatment facilities, and other factors. However, the utility of ZLD may be particularly acute for inland desalination locations, which often have fewer alternatives than coastal regions. For example:

- An inland area may lack a suitable receiving body for desalination concentrate, or the concentrate water quality may not permit discharge to the receiving bodies that exist.
- There may be a lack of centralized wastewater collection system for discharge, or if such infrastructure is present, the municipal effluent discharge permit may not allow for higher salinity flows.
- Water quality and/or geology may limit deep-well injection.
- The evaporation rate and/or availability of land acreage for storage ponds may be insufficient to accommodate the concentrate flow.

If all of these conditions are true, then ZLD may be the only alternative for concentrate management and thus the only means of enabling the use of desalination treatment. Unlike most other concentrate management strategies, "conventional" ZLD (i.e., as defined previously) can be deployed almost anywhere and has relatively few environmental, regulatory, or practical limitations that restrict feasibility; however, the cost may be prohibitive.

Zero Liquid Discharge Treatment

ZLD treatment may be applied to the concentrate of a primary desalination process itself or to the residuals from an enhanced recovery step, such as those discussed previously. In the latter case, the flow directed to the ZLD process would be smaller and more highly concentrated, reducing the cost; however, the ZLD cost savings must be evaluated against the additional capital and operating cost of the enhanced recovery process on a case-specific basis.

Conceptually, ZLD can be considered as a series of two progressive residuals treatment steps: advanced concentration followed by solids generation. Although these two steps may not necessarily be distinct, depending on the specific processes used, the underlying concept universally applies and is useful for understanding ZLD treatment. Accordingly, the discussion of ZLD treatment in this chapter is focused on the combination of brine concentrators (advanced concentration) and crystallizers (solids generation). While there may be many varied approaches to achieving ZLD, the use of brine concentrators and crystallizers has the important advantage of not being limited by water quality. For example, Locke et al. (2015) describe a groundwater desalination plant owned by the City of Palm Coast, Fla., that achieves ZLD by applying a combination of precipitative softening, ultrafiltration (UF), sludge thickening, and dewatering to its RO concentrate. (A case study of this facility is provided in Chapter 8.) However, the successful implementation of this technique is predicated on a raw water supply with relatively low TDS levels composed predominantly of hardness; this same approach would not be suitable for the concentrate of a seawater desalination plant consisting largely of sodium and chloride, which do not readily precipitate. By contrast, the coupling of brine concentrators and crystallizers can accomplish ZLD with almost any saline supply, independent of TDS concentrations or the relative proportion of its constituent dissolved solids.

Descriptions of both brine concentrators and crystallizers are provided subsequently, as compiled from various corroborating references, including Ciszewski 2015, Yallaly 2015, Juby et al. 2008, and Mickley 2006. Because several of these sources, as well as others in industry literature, offer detailed process discussions of brine concentrators and

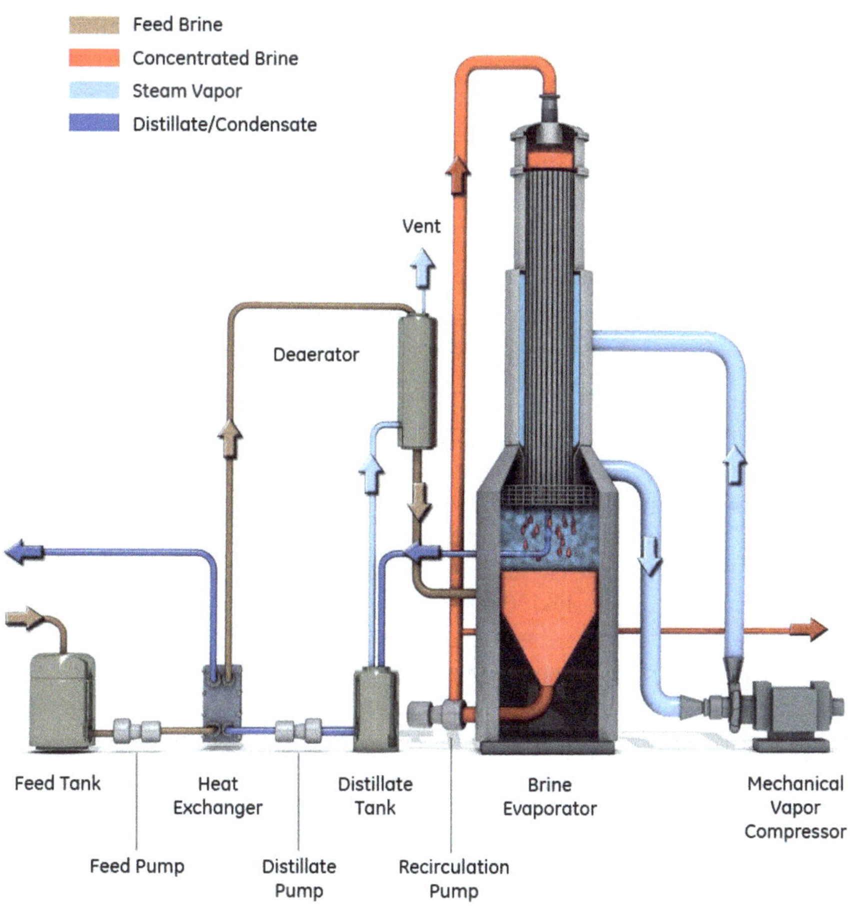

Courtesy of GE Water & Process Technologies

Figure 5-14 Brine concentrator process schematic (typical)

crystallizers, such information is not repeated in this manual. Instead, the purpose of the descriptions that follow is to provide a general overview and understanding of brine concentrator and crystallizer technology, focusing on the points most critical for a planning-level perspective in a concise summary format that is not available elsewhere. In accordance with the mandate of this manual, special emphasis is given to considerations that are most relevant to inland desalination applications. Although the descriptions that follow are framed in the context of employing RO as a primary desalination process, this assumption does not necessarily limit the discussion from applying to the use of brine concentrators and crystallizers with the concentrate of other desalination technologies.

Brine concentrators. Unlike RO, which employs pressure-driven membrane separation to accomplish desalination, brine concentrators use mechanical vapor compression technology to desalinate via thermal evaporation. Figure 5-14 shows a process schematic for a typical brine concentrator.

For an application to treat desalination plant residuals, the influent "feed brine" shown in Figure 5-14 represents the concentrate from either a primary RO process or an enhanced recovery process applied to the primary RO concentrate as an intermediate step. Acid may be added to the brine concentrator feed to lower the pH, thereby converting bicarbonate to dissolved carbon dioxide, which is subsequently removed in the deaerator to prevent corrosion. Because scaling onto heat transfer surfaces can reduce efficiency, several scale control measures may be implemented in conjunction with brine concentrator operation.

- It is common to seed calcium sulfate crystals into the recirculated brine slurry (shown in Figure 5-14 as the red "concentrated brine" pipe loop driven by the recirculation pump) to serve as precipitation nuclei, preventing scaling by facilitating crystal growth.
- Scale-inhibiting chemicals are routinely added to the feed to control silica and other sparingly soluble species.
- The deaeration process that removes dissolved carbon dioxide from the system provides the dual benefit of removing alkalinity, thus reducing the potential for calcium carbonate scaling.

The "distillate/condensate" flow in Figure 5-14 is the treated water produced by the brine concentrator, and TDS concentrations are typically below 10 mg/L. If a brine concentrator is used as the first step in a ZLD process (as is the case in this discussion), then the bleed flow from the "concentrated brine" shown in Figure 5-14 (red pipe and arrow at the bottom right) serves as the crystallizer feed flow. TDS levels in this residuals stream generally vary from approximately 180,000 to 250,000 mg/L, depending on the scaling potential of the constituent dissolved solids; concentrations can potentially be increased to about 300,000 mg/L if a softening pretreatment process is used.

Water recovery across the brine concentrator (i.e., the percent of influent feed flow that is converted to distillate) varies with the feed TDS concentration and its associated scaling potential. These influences are demonstrated in Table 5-1, which shows the brine concentrator recovery at feed TDS levels ranging from 5,000 to 50,000 mg/L for each of three cases in which the effluent "concentrated brine" (as referenced in Figure 5-14) is limited to 200,000, 250,000, and 300,000 mg/L, respectively, because of scaling. As indicated by these example cases, feedwater with lower TDS and less scaling potential (i.e., higher TDS limits in the concentrated brine effluent) results in higher brine concentrator recoveries. Although this latter phenomenon may seem counterintuitive, it is governed by a simple mass balance across the brine concentrator in which both the brine and distillate concentrations

Table 5-1 Influence of feed TDS and maximum concentrated brine effluent TDS on brine concentrator recovery[1,2]

Brine Concentrator Feed TDS (mg/L)	Brine Concentrator Recovery at a Maximum[3] Concentrated Brine Effluent TDS of		
	200,000 mg/L	250,000 mg/L	300,000 mg/L
5,000	97.5%	98.0%	98.3%
10,000	95.0%	96.0%	96.7%
15,000	92.5%	94.0%	95.0%
20,000	90.0%	92.0%	93.3%
25,000	87.5%	90.0%	91.7%
30,000	85.0%	88.0%	90.0%
35,000	82.5%	86.0%	88.3%
40,000	80.0%	84.0%	86.7%
45,000	77.5%	82.0%	85.0%
50,000	75.0%	80.0%	83.3%

[1] Assumes brine concentrator distillate TDS of 10 mg/L.
[2] Brine concentrator recoveries cited in the table are independent of feed flow.
[3] As limited by assumed scaling potential.

are fixed. (Distillate TDS levels will exhibit small variations but are relatively stable; for the examples provided in Table 5-1, the distillate TDS was assumed to be 10 mg/L.) Moreover, because the TDS levels for the three flows (i.e., feed, brine, and distillate) are fixed for each case shown in Table 5-1, the relative proportions of the flows are likewise constant; consequently, for each case, the indicated recovery is independent of the flows, such that any set of flow values that satisfy the mass balance will yield the same recovery. Note that the recoveries in Table 5-1 are for the brine concentrator process only; if a brine concentrator and crystallizer are used as successive steps in a ZLD treatment process, the overall combined recovery is assumed to be effectively 100 percent (i.e., insignificant volume of liquid residuals).

Although the context of this discussion is the combination of brine concentrators and crystallizers as "conventional" ZLD that can be applied largely independent of feedwater chemistry and flow, the use of a crystallizer is not always prerequisite for producing crystalline solids for disposal. If the flow of concentrated brine from the brine concentrator is low enough, then discharge to a lined evaporation pond may be feasible (potential limitations of land area and evaporation rate notwithstanding). A case study of a California facility using this approach is provided in Chapter 8.

Brine concentrator–specific energy consumption ranges from about 60 to 90 kWh per 1,000 gallons (kgal) of water recovered (i.e., treated water flow), with higher degrees of concentration (e.g., targets of 200,000 mg/L vs. 250,000 mg/L vs. 300,000 mg/L, etc.) requiring more energy per unit volume (Ciszewski 2015, Juby et al. 2008, Mickley 2006). By contrast, seawater desalination using RO (SWRO) typically requires 10–15 kWh/kgal of permeate. SWRO and brine concentrators are not functionally equivalent; the former is a primary desalination process that produces concentrate on the order of about 70,000 mg/L, while the latter would be applied to this SWRO concentrate and yield a residuals stream with TDS levels roughly two to four times higher. This comparison is provided only to illustrate the large increase in specific energy necessary to achieve concentrated solutions significantly higher than those produced by a primary desalination process. Note that the values for brine concentrator–specific energy consumption provided in this discussion assume the use of electric power and do not reflect the use of any available waste heat or other thermal advantage; consequently, the range cited previously represents a conservative estimate that is appropriate for planning purposes.

Aside from the high energy requirements, the disadvantages of brine concentrators are primarily associated with operations and maintenance. In addition to the system complexity, brine concentrators respond more slowly to flow changes than a primary RO process, thus necessitating appropriately sized equalization tanks for the feed flow (as illustrated in Figure 5-14). Moreover, a supply of steam is necessary for brine concentrator startup, including after every maintenance event, both complicating and extending the duration of the startup process relative to many other treatment technologies; this steam may be provided by a dedicated boiler or supplied by another steam-generating process on site (e.g., in the case of co-location with an applicable industrial manufacturing or power facility).

These challenges can be especially problematic for potable water treatment system operators, who may not have experience with the complex thermal desalination systems that are uncommon in municipal applications outside of the Middle East. Thus, a rigorous operator training program is essential. In addition, because maintenance can be time- and labor-intensive, design redundancy is often necessary to ensure uninterrupted treatment capacity. Aesthetic concerns may also be a drawback, depending on the location of the treatment facility. Because brine concentrators (and crystallizers) bear more resemblance to industrial equipment than the processes used at many drinking water treatment plants, public acceptance may be an important consideration. Figure 5-15 shows a facility employing both crystallizer (one unit at left) and brine concentrator (two units at right) processes.

Courtesy of GE Water & Process Technologies

Figure 5-15 Crystallizer (left) and brine concentrators (typical)

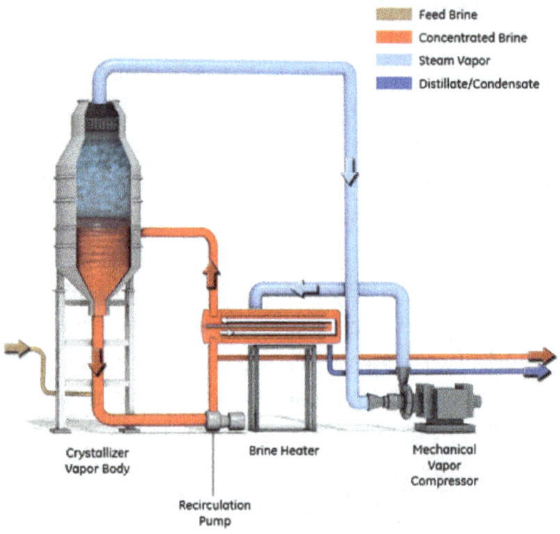

Courtesy of GE Water & Process Technologies

Figure 5-16 Crystallizer process schematic (typical)

Crystallizers. Like brine concentrators, crystallizers use mechanical vapor compression to desalinate highly concentrated brines, generating a slurry of solids. Figure 5-16 shows a process schematic for a typical brine concentrator.

In a conventional ZLD treatment scheme (i.e., the combination of brine concentrators and crystallizers), the "Feed Brine" shown in Figure 5-16 is the concentrated residuals stream produced by the brine concentrator. Because the primary purpose of the crystallizer is to produce precipitated solids, neither acid nor scale inhibitors are used for pretreatment; however, antifoaming agents may be added. Concentrate from a primary RO process could also be treated with a crystallizer, although the costs would be much

Table 5-2 Influence of flow on crystallizer operating costs (example)

Feedwater	Crystallizer Specific Energy Requirement kWh/kgal	Crystallizer Feed Flow mgd	Annual Energy Usage[2] 10^6 kWh	Energy Cost[3] $
Brine concentrator residuals[1]	200	0.1	7.3	$730,000
Primary reverse osmosis concentrate		1	73	$7,300,000

[1] Concentrated brine, as depicted in Figure 5-16
[2] Assumes continuous operation
[3] Assumes $0.10/kWh

greater because of the extremely high specific energy requirements, which can range from about 180 to 250 kWh/kgal (Ciszewski 2015, Mickley 2006). Table 5-2 shows an example of the influence of the higher RO concentrate flows (versus brine concentrator residuals flows) on crystallizer operating costs. The higher flows associated with RO concentrate would also substantially affect the capital cost of a crystallizer process, as the required capacity would be approximately an order of magnitude higher than would be the case as applied to the brine concentrator residuals stream.

Water quality can also have an influence on the specific energy requirements for crystallizers. If the feedwater has high concentrations of species that more readily precipitate out of solution, the specific energy requirements will be lower; however, significant concentrations of highly soluble species such as nitrate salts require more energy per unit volume of distillate produced. Also, as discussed in conjunction with brine concentrators, the 180–250 kWh/kgal range for crystallizer-specific energy consumption provided in this discussion is a conservative estimate for planning purposes that assumes the use of electric power and does not reflect the use of any available waste heat or other thermal advantage.

The distillate/condensate flow in Figure 5-16 is the treated water produced by the crystallizer, and TDS concentrations are typically about 30–50 mg/L, a range that is somewhat higher than brine concentrator distillate (Ciszewski 2015). The recovery of the crystallizer process is effectively 100 percent (i.e., the volume of liquid waste is insignificant), although this includes the use of a centrifuge to dewater the solid slurry residual stream (i.e., the concentrated brine in Figure 5-16) to produce crystalline solids for landfilling or beneficial use (if feasible). The centrate generated by centrifuge is typically returned to the crystallizer feed flow. (The centrifuge and centrate recycle are not shown in Figure 5-16.)

Crystallizers have many of the same disadvantages as brine concentrators: operational complexity, the need for a supply of steam for each system startup, and aesthetics. Crystallizers also respond relatively slowly to flow changes; although it is not shown in Figure 5-16, equalization storage is likewise recommended for the crystallizer feed flow.

Residuals Management

Although a true ZLD treatment scheme does not generate liquid residuals (by definition), the process does produce solids (or a highly concentrated solid slurry) that must be properly managed for either beneficial use or disposal. Because these solids may contain significant levels of regulated metals or other contaminants, it is important to conduct a hazardous waste characterization as a component of any management strategy (as discussed in Chapter 7). Solids that cannot be recycled for beneficial use should be directed to a landfill (or similar facility) that is appropriately permitted to receive the residuals in accordance with their characterization. ZLD residuals are typically transported by truck, and the cost of this transportation, as well as the comprehensive cost of residuals management, should be incorporated into any economic analysis of ZLD. Note that any economic advantage from beneficial use would also be a factor in this analysis. A broader discussion of residuals management in association with inland desalination is provided in Chapter 9 of this manual.

Cost Considerations

Critical considerations regarding the cost of "conventional" ZLD treatment are discussed in the following subsections, including appropriate evaluation and normalization, along with factors that influence ZLD costs. Note that this section is not intended to overlap with the broad presentation of inland desalination costs provided in Chapter 6 of this manual, but rather to serve as a complementary discussion that conveys important details specific to ZLD systems.

Cost comparison and normalization. As discussed previously, the specific energy consumption for both brine concentrators and crystallizers is much higher than that for seawater desalination, which is among the most energy-intensive processes used in potable water treatment. These substantial energy requirements translate into very high ZLD treatment costs. Examples of ZLD costs for systems with four different capacities are summarized in Table 5-3, which is adapted from a 2008 USBR study (Report No. 149 in the series prepared under the Desalination and Water Purification Research and Development Program; Juby et al. 2008). The costs in Table 5-3 include both brine concentrators and crystallizers, and 100 percent recovery is assumed, such that the feed and treated water flows are equivalent; capital costs were amortized over 20 years at a 6 percent annual interest rate. Note that the ZLD costs in USBR Report No. 149 include landfilling of solids at $50/ton; this cost could increase if testing determines that the crystallized solids constitute a hazardous waste.

As shown in Table 5-3, amortized ZLD unit costs range from approximately $30/kgal to $60/kgal over capacities spanning from 1 mgd down to 0.125 mgd. For the purposes of comparison, AWWA M61 (*Desalination of Seawater*) cites amortized unit costs for SWRO that vary from about $2.50/kgal to $5/kgal over a wide range of capacities (2011). Even at unit costs that are double these figures (i.e., $5/kgal–$10/kgal), SWRO—one of the most expensive water treatment processes—is much less expensive than ZLD.

The difference in costs is even more pronounced for the desalination of brackish water, which is more characteristic of inland locations, as shown in Table 5-4. The cost examples in this table are adapted from Bergman and Elarde (2003), who developed best-fit curves of cost figures from operating desalination plants. For the purposes of comparison with the ZLD costs published in USBR Report No. 149, the costs in Table 5-4 are escalated to 2008 dollars (as applicable) and amortized using similar terms (i.e., 20 years at 6 percent

Table 5-3 ZLD treatment cost example[1]

	Capacity[2]	Capital Cost		O&M Cost		Total Amortized Cost[3]	
Case	mgd	$M	$/gpd	$M/yr	$/kgal	$M/yr	$/kgal
ZLD-1	1	$41.11	$41.11	$7.05	$19.11	$10.63	**$29.12**
ZLD-2	0.5	$28.38	$56.77	$3.78	$20.29	$6.25	**$34.25**
ZLD-3	0.25	$21.53	$86.13	$2.16	$23.67	$4.04	**$44.24**
ZLD-4	0.125	$16.64	$133.11	$1.28	$28.07	$2.73	**$59.87**

[1] Juby et al. 2008 (for "Brine B")
[2] Assumes 100% recovery (i.e., ZLD feed flow = ZLD treated water flow)
[3] Amortization over 20 years at a 6% annual interest rate

Table 5-4 RO treatment cost example at 80% recovery

	Capacity					
	Feed	Permeate	Concentrate	Capital Cost[1]	O&M Cost[1]	Total Amortized Cost[2]
Case	mgd	mgd	mgd	$/gpd	$/kgal	$/kgal
RO-1	5	4	1	$0.58	$0.30	**$0.44**
RO-2	2.5	2	0.5	$0.80	$0.40	**$0.59**
RO-3	1.25	1	0.25	$1.11	$0.60	**$0.87**
RO-4	0.625	0.5	0.125	$1.54	$0.70	**$1.07**

[1] Adapted from Bergman and Elarde 2003 (escalated to 2008 dollars)
[2] Amortization over 20 years at a 6% annual interest rate

Table 5-5 RO+ZLD combined treatment cost example at 80% RO recovery

	Capacity[1]	Capital Cost	O&M Cost	Total Amortized Cost[2]
Case	mgd	$/gpd	$/kgal	$/kgal
RO-1 ZLD-1	5	$8.68	$4.10	**$6.18**
RO-2 ZLD-2	2.5	$11.99	$4.46	**$7.32**
RO-3 ZLD-3	1.25	$18.11	$5.21	**$9.54**
RO-4 ZLD-4	0.625	$27.85	$6.17	**$12.83**

[1] Assumes 100% recovery over the treatment plant (i.e., feed flow = RO permeate flow + ZLD distillate flow)
[2] Amortization over 20 years at a 6% annual interest rate

interest). The cases shown in Table 5-4 are designed to align with those in Table 5-3, such that the concentrate of case RO-1 becomes the feed to the ZLD-1 case. A recovery of 80 percent was used for the example RO systems, which were assumed to be treating brackish water with a TDS concentration similar to the raw water supply used to generate the ZLD costs in USBR Report No. 149: ~2,300 mg/L treated with RO, with the concentrate feeding the ZLD process.

A comparison of the amortized unit costs in Tables 5-3 and 5-4 indicates that ZLD treatment is roughly one to two orders of magnitude more expensive than typical brackish water desalination using RO. Note that the figures in Table 5-4 are not intended as a precise estimate of RO costs but rather are representative figures for the purposes of illustration; even significant but reasonable variation of the total amortized costs for RO in Table 5-4 will not render these costs comparable to those for ZLD shown in Table 5-3.

Cost comparisons such as these are common in literature discussing ZLD, and they are valid for illustrating the starkly higher expense of ZLD treatment. However, this exercise is not a truly equitable comparison given that the unit costs for ZLD and RO are normalized for the flows treated by each process: distillate and permeate, respectively. In practice, an inland desalination facility employing a ZLD concentrate management strategy would have both RO and ZLD processes, and the total cost of treatment (i.e., RO+ZLD) should be normalized by the total treated water flow (i.e., RO permeate + ZLD distillate). The impact of properly accounting for the capital, operating, and total amortized costs of a combined RO (from the examples in Table 5-4) and ZLD (from the examples in Table 5-3) is illustrated in Table 5-5.

The total amortized costs of RO (from Table 5-4), ZLD (from Table 5-3), and combined RO+ZLD (Table 5-5) systems are summarized in Table 5-6 for direct comparison. Although the total amortized costs for the combined treatment are generally an order of magnitude higher than for RO alone, they are also roughly an order of magnitude less expensive than ZLD alone.

Table 5-6 Sample RO and ZLD treatment cost comparison at 80% RO recovery

	RO System		ZLD System		Combined Treatment	
	Permeate Flow	Total Amortized Cost[1]	Distillate Flow	Total Amortized Cost[1]	Treated Water Production	Total Amortized Cost[1]
Case	mgd	$/kgal	mgd	$/kgal	mgd	$/kgal
RO-1 → ZLD-1	4	**$0.44**	1	**$29.12**	5	**$6.18**
RO-2 → ZLD-2	2	**$0.59**	0.5	**$34.25**	2.5	**$7.32**
RO-3 → ZLD-3	1	**$0.87**	0.25	**$44.24**	1.25	**$9.54**
RO-4 → ZLD-4	0.5	**$1.07**	0.125	**$59.87**	0.625	**$12.83**

mgd—million gallons per day, RO—reverse osmosis, ZLD—zero liquid discharge
[1] Amortization over 20 years at a 6% annual interest rate

Table 5-7 RO treatment cost example at 90% recovery

	Capacity					
	Feed	Permeate	Concentrate	Capital Cost[1]	O&M Cost[1]	Total Amortized Cost[2]
Case	mgd	mgd	mgd	$/gpd	$/kgal	$/kgal
RO-1A	5	4.5	0.5	$0.55	$0.25	**$0.38**
RO-2A	2.5	2.25	0.25	$0.76	$0.35	**$0.53**
RO-3A	1.25	1.125	0.125	$1.05	$0.60	**$0.85**

mgd—million gallons per day, O&M—operation and maintenance, RO—reverse osmosis
[1] Adapted from Bergman and Elarde 2003 (escalated to 2008 dollars)
[2] Amortization over 20 years at a 6% annual interest rate

Table 5-8 RO+ZLD combined treatment cost example at 90% RO recovery

	Capacity[1]	Capital Cost	O&M Cost	Total Amortized Cost[2]
Case	mgd	$/gpd	$/kgal	$/kgal
RO-1A → ZLD-2	5	$6.17	$2.29	**$3.77**
RO-2A → ZLD-3	2.5	$9.30	$2.68	**$4.90**
RO-3A → ZLD-4	1.25	$14.26	$3.35	**$6.75**

mgd—million gallons per day, O&M—operation and maintenance, RO—reverse osmosis, ZLD—zero liquid discharge
[1] Assumes 100% recovery over the treatment plant (i.e., feed flow = RO permeate flow + ZLD distillate flow)
[2] Amortization over 20 years at a 6% annual interest rate

Table 5-9 Sample RO and ZLD treatment cost comparison at 90% RO recovery

	RO System		ZLD System		Combined Treatment	
	Permeate Flow	Total Amortized Cost[1]	Distillate Flow	Total Amortized Cost[1]	Treated Water Production	Total Amortized Cost[1]
Case	mgd	$/kgal	mgd	$/kgal	mgd	$/kgal
RO-1A → ZLD-2	4.5	**$0.38**	0.5	**$34.25**	5	**$3.77**
RO-2A → ZLD-3	2.25	**$0.53**	0.25	**$44.24**	2.5	**$4.90**
RO-3A → ZLD-4	1.125	**$0.85**	0.125	**$59.87**	1.25	**$6.75**

mgd—million gallons per day, RO—reverse osmosis, ZLD—zero liquid discharge
[1] Amortization over 20 years at a 6% annual interest rate

Because minimizing the flow to the ZLD process can significantly reduce treatment costs, the impact of increasing RO recovery in a combined RO+ZLD system can have a substantial financial impact. Tables 5-7, 5-8, and 5-9 replicate Tables 5-4, 5-5, and 5-6, respectively, assuming the RO recovery has been increased from 80 to 90 percent. Thus, the RO-1 system concentrate flow changes from 1 to 0.5 mgd (case RO-1A), which then aligns with ZLD flow in case ZLD-2. The RO-2 concentrate flow decreases from 0.5 to 0.25 mgd (case RO-2A), which aligns with the ZLD flow in case ZLD-3. Likewise, the RO-3 concentrate is reduced from 0.25 to 0.125 mgd (case RO-3A), aligning with the ZLD flow in case ZLD-4. (Because the cost data provided in USBR Report No. 149 are for discrete and specific flows, there is no companion ZLD example for the case in which the RO-4 system concentrate decreases by half.)

Note that this example of increasing the RO recovery to 90 percent neglects the resulting change in concentrate salinity. Although the TDS levels in the RO concentrate would roughly double in cases in which the recovery was increased from 80 to 90 percent (i.e., halving the concentrate flow while assuming a constant 100 percent salt rejection for the purposes of illustration), the impact on ZLD treatment costs because of this change in salinity would be very small relative to the influence of the decreased flow. However, both the decreased flow and increased TDS would result in lower ZLD treatment costs.

The influence of increasing the recovery of a primary RO process from 80 to 90 percent is summarized in Table 5-10, including the respective total amortized costs and the percent cost reduction associated with the recovery increase. Note that these examples are not representative of installed facilities making adjustments to increase RO recovery from 80 to 90 percent; in these cases, the capacity of the ZLD system would be oversized by a factor of two, skewing the capital costs. Instead, the example in Table 5-10 illustrates the comparison of options in the planning stage and the impact on total amortized cost if 90 percent versus 80 percent primary RO recovery can be achieved in conjunction with a ZLD system sized appropriately for each case. As Table 5-10 shows, the impact of increased recovery is substantial, reducing the total amortized unit cost of treated water by about 30–40 percent for the examples shown.

Overall, this analysis makes the following three points about "conventional" ZLD systems (i.e., the combination of brine concentrators and crystallizers).

1. Economies of scale are significant.

As illustrated in Table 5-3, economies of scale significantly affect the unit cost of water produced by a ZLD system. This impact likewise heavily influences the unit cost of desalinated water in a combined RO+ZLD system as summarized in Table 5-10. Thus, to the extent that larger-capacity systems are feasible and affordable (i.e., within budget), it is economically advantageous to maximize the size of a planned RO+ZLD desalination plant.

Table 5-10 Comparison of RO+ZLD combined treatment costs at 80% versus 90% RO recovery

	RO+ZLD @ 80% RO Recovery		RO+ZLD @ 90% RO Recovery		
Treated Water Capacity[1]	Case	Total Amortized Cost[2] $/kgal	Case	Total Amortized Cost[2] $/kgal	Percent Unit Cost Reduction
5	RO-1 → ZLD-1	**$6.18**	RO-1A → ZLD-2	**$3.77**	39%
2.5	RO-2 → ZLD-2	**$7.32**	RO-2A → ZLD-3	**$4.90**	33%
1.25	RO-3 → ZLD-3	**$9.54**	RO-3A → ZLD-4	**$6.75**	29%

RO—reverse osmosis, ZLD—zero liquid discharge
[1] Assumes 100% recovery over the treatment plant (i.e., feed flow = RO permeate flow + ZLD distillate flow)
[2] Amortization over 20 years at a 6% annual interest rate

2. Maximizing primary RO recovery is important for ZLD affordability.

For a given total available flow of water (e.g., the rights to 5 mgd of a brackish water supply), it is economically advantageous to maximize primary RO permeate flow (e.g., 4.5 mgd versus 4 mgd) and minimize the flow to the ZLD process (e.g., 0.5 mgd versus 1 mgd). Applying an enhanced recovery process, such as those described previously in this chapter, to the RO concentrate could likewise reduce the ZLD process flow, although the capital and operating costs of the enhanced recovery process must also be taken into account in determining the total amortized unit cost of the treated water.

3. The unit costs of seawater desalination and inland desalination with ZLD treatment are roughly comparable.

The properly amortized unit costs of RO+ZLD treatment summarized in Table 5-10 are on the same order of magnitude as the $2.50/kgal–$5/kgal figures indicated for SWRO in AWWA M61 (*Desalination of Seawater*), and they compare very well for the RO+ZLD cases at 90 percent RO recovery. Accounting for the RO+ZLD system economies of scale, as demonstrated by the declining amortized unit cost trends with increasing treated water capacity shown in Table 5-10, larger RO+ZLD systems would likely compare even more favorably with SWRO. Thus, although SWRO is not an option for inland water suppliers, these utilities do have the economically comparable alternative of brackish water desalination in combination with ZLD treatment. The latter case is limited only by the accessibility of brackish water supplies, not by either prohibitively disproportionate economics or the lack of a viable concentrate management strategy.

Other cost considerations. Because of the large specific energy consumption (kWh/kgal) associated with both brine concentrators and crystallizers, the unit price of energy ($/kWh) exerts a substantial influence on the cost of desalination for cases in which "conventional" ZLD treatment is used for concentrate management. Likewise, for these same cases, because of the high unit cost of treating water using ZLD (expressed as $/kgal in Table 5-3), minimizing the ZLD flow can significantly reduce the overall cost of desalination. Thus, as applied to a primary RO process, minimizing the concentrate (i.e., ZLD feed) flow reduces both the required ZLD capacity and the associated energy expended, resulting in lower capital and operating costs, respectively.

Although ZLD capital costs are not sensitive to changes in water quality, the operating cost is influenced by the concentrations of both TDS and relative levels of the constituent components of the TDS. For example, in comparing two cases in which ZLD feed flows are similar, the flow with the higher TDS requires less concentration, reducing the specific energy consumption, which in turn lowers the cost. In terms of constituent dissolved solids, greater proportions of highly soluble species will increase the specific energy necessary to accomplish ZLD treatment; nitrates are among the most notable of these soluble species.

The economy of scale associated with ZLD treatment is demonstrated in Table 5-3, with higher unit costs for smaller capacity systems. Thus, in terms of the cost per unit volume of water produced, larger ZLD systems are more economical than small facilities. However, the economy of scale is limited by the size of discrete brine concentrator and crystallizer units for which components can be transported; currently, this limit is achieved for ZLD systems with a capacity in the range of about 1,000–1,200 gpm (Ciszewski 2015). Beyond this limit, larger-capacity ZLD systems consist of increasing numbers of discrete treatment units (i.e., rather than larger units), negating further economies of scale.

Installations and Institutional Knowledge

Although ZLD treatment is used in many industrial applications, at the time of this manual there is only one known municipal installation in the United States using a brine concentrator to treat the residuals from a primary RO process, and the resulting concentrated brine is directed to a lined evaporation pond. There are no known municipal applications of crystallizer treatment. Thus, institutional knowledge regarding ZLD remains extremely limited in the potable water and reuse sectors. However, despite the dearth of municipal installations, the two processes described as "conventional" ZLD in this manual (i.e., brine concentrators and crystallizers) are proven technologies that are well understood by experienced professionals. Consequently, utilities evaluating the implementation of ZLD may opt to rely on performance-based equipment specifications rather than detailed mechanical and process design conducted by a third party. In addition, because of the complexity of ZLD systems, the use of contract operation may also be a strategic consideration.

REFERENCES

AWWA. 2011 (1st ed.). M61, *Desalination of Seawater*. Denver: AWWA.

AWWA. 1995. M38, *Electrodialysis and Electrodialysis Reversal*. Denver: AWWA.

Benjamin, M.; Shi, W.; Kwan, P.; and Chang, Y. 2010. Evaluation of VSEP to Enhance Water Recovery During Treatment of Brackish Water and RO Concentrate. Denver: Water Research Foundation.

Bergman, B. and Elarde, J. 2003. The Cost of Membrane Softening and Desalting for Municipal Water Supplies. Proc. AWWA Membrane Technology Conf., Atlanta, Ga.

Ciszewski, D. 2015. Personal communication.

Elimelech, M. 2014. Forward Osmosis: Progress and Challenges. Proc. 2014 National Water Research Institute Clarke Prize Conf., Huntington Beach, Calif.

Faust, S.A. and Aly, O.M. 1998 (2nd ed.). *Chemistry of Water Treatment*. Boca Raton, Fla.: CRC Press.

Federal Efficiency Management Program (FEMP). 2004. *Saving Energy, Water, and Money with Efficient Water Treatment Technologies*. https://www.nrel.gov/docs/fy04osti/34721.pdf (accessed Dec. 3, 2018).

General Electric (GE). 2010. Modular Ultrapure Water Plant—Fact Sheet. Boston: GE.

Giesen, A. 1999. Crystallisation Process Enables Environmentally Friendly Phosphate Removal at Low Costs. *Environmental Technology*, 20(7):769–775.

Juby, G.; Zacheis, A.; Shih, W.; Ravishanker, R.; Mortazavi, B.; and Nusser, M. 2008. Evaluation and Selection of Available Processes for a Zero-liquid Discharge System for the Perris, California, Ground Water Basin. Desalination and Water Purification Research and Development Program Report No. 149. Washington: USBR.

Juby, G.J.G.; Schutte, C.F.; and Van Leeuwen, J. 1996. Desalination of Calcium Sulphate Scaling Mine Water: Design and Operation of the SPARRO Process. *Water SA*, 22(2):161–172.

Locke, P.; Greiner, F.; Popko, R.; and Cooke, N. 2015. ZLD Process Utilizes Ultrafiltration to Achieve 98.5% RO Recovery. Proc. 2015 AMTA/AWWA Membrane Technology Conf., Orlando, Fla.

Mansell, B.; Ackman, P.; Tang, C.; and Friess, P. 2014. Pilot-Scale Testing of the Precipitative Softening-Secondary RO Process to Reduce RO Concentrate Volume. Proc. Water Environment Federation Technical Exhibition and Conf., New Orleans, La.

Mickley, M. 2006 (2nd ed.). Membrane Concentrate Disposal: Practices and Regulation. Desalination and Water Purification Research and Development Program Report No. 123. Denver: USBR.

O'Neail, T.M.; Kirchner, O.E.; and Day, W.J. 1981. Achieving High Recovery From Brackish Water With Seeded Reverse Osmosis Systems. Proc. 42nd International Water Conference, Pittsburgh, Pa.

USBR. 2010. Lime Softening Fact Sheet. Washington: USBR.

Water Desalination Report. 2012. *Oasys Membrane Brine Concentrator (MBC) Process*. 48:10.

Wiesner, A.D.; Juby, G.J.G.; and Ross, N. 2015. Combining EDR and SPARRO Technologies to Increase Recovery at Inland Desalters. Proc. 2015 AWWA Annual Conf. and Exposition, Anaheim, Calif.

Yallaly, B. 2015. Personal communication.

Chapter 6

Cost of Desalination and Concentrate Management

Jorge T. Aguinaldo and Rick Bond

This chapter covers considerations and approaches for estimating the treatment cost for inland desalination and concentrate management. Even though most inland desalters use the same desalination process, RO, the cost for desalinating brackish water and managing the resulting concentrate byproduct is highly site-specific. Many local factors including water quality, treatment goals, process selection, concentrate management options, climate, and land availability can play a significant role in determining treatment cost. Furthermore, concentrate management is a young field that is evolving rapidly as new ideas and technologies are developed in an effort to reduce costs and energy consumption. Consequently, the treatment cost for concentrate management is evolving. Treatment costs for some of the more established concentrate methods such as evaporation ponds and brine concentrators are included in this chapter, but emerging technologies without established costs are not.

As a general rule, desalinating brackish water and managing the concentrate is more expensive and more variable than treating freshwater. Therefore, development of cost estimates early in the planning stages of the project is critical to selecting the optimal treatment approach, and in many cases, cost may dictate the viability of an inland desalination project.

It is not advisable to try to obtain accurate site-specific treatment costs for inland desalination and concentrate management directly from a manual or general guidance documents. Rather, treatment costs must be developed for each site using a method appropriate for the level of accuracy required. The objective of this chapter is to provide a framework for developing site-specific treatment costs for inland desalination and concentrate management appropriate for concept screening and feasibility studies.

OVERALL COST CONSIDERATIONS

Total treatment cost comprises all of the costs required to design, build, and operate the treatment plant. Treatment cost is often expressed as a cost per unit of product water such as dollars per 1,000 gallons (kilo-gallon): $/kgal. Cost per unit of water produced is calculated by amortizing the construction cost into an annual debt-service cost, adding to it all of the annual costs, and dividing the sum of these two costs by the desalination plant freshwater production capacity and the design availability factor of the plant. The design availability factor is the percentage of time per year for which the plant is designed to produce freshwater flow greater than or equal to the design plant capacity; usually such factor is 90–96 percent.

The total cost to build treatment facilities includes construction (direct capital) and nonconstruction (indirect capital) project costs. Construction costs are the direct costs incurred in the construction of all plant facilities, including materials, labor, and equipment. Nonconstruction project costs include expenses associated with the development and implementation activities of the project, such as engineering, legal, land acquisition, contingencies, permitting costs and associated environmental studies and mitigation, and interest during construction. Annual expenses include costs for operation and maintenance (O&M), energy, labor, chemicals, debt service, and administration and management of plant operations.

Table 6-1 summarizes items typically included in each category.

The level of effort and detail used to develop a cost estimate should depend on usage of the cost estimate. For example, cost estimates developed for budget authorization should be more detailed and accurate than cost estimates developed for process screening.

Table 6-2 shows guidelines for cost estimate classes recommended by the Association for the Advancement of Cost Engineering (AACE). Low and high ranges for accuracy are listed for each class. This accuracy range narrows from Class 5 to Class 1.

Table 6-1 Summary of major treatment cost by category

Total Project Cost Items		Annual Project Cost Items
Direct Capital or Construction Costs	Indirect Capital or Nonconstruction Costs	Annual Project Costs
• Project equipment supply and installation • Well fields and/or intakes • Pump stations • Pipelines and outfalls • Water storage tanks • Waste disposal facilities • Power supply	• Engineering • Construction management • Permitting, environmental studies, and mitigation • Legal and financing • Interest during construction • Contingencies	• Debt service • Operation and maintenance • Energy • Labor • Chemicals • Membrane replacement • Administration and management of plant operations • Permit compliance monitoring and reporting

Table 6-2 Guidelines for cost estimate classes

Estimate Class	End Usage	Typical Variation in Accuracy Range
Class 5	Concept screening	Low: –20% to –50% High: +30% to +100%
Class 4	Study or feasibility	Low: –15% to –30% High: +20% to +50%
Class 3	Budget authorization	Low: –10% to –20% High: +10% to +30%
Class 2	Bid	Low: –5% to –15% High: +5% to +20%
Class 1	Check estimate	Low: –3% to –10% High: +3% to +15%

Source: Adapted from AACE 2005

The treatment costs often are developed based on material or equipment costs from previous projects of similar size, intake, outfall, and desalination technology. When using cost data from an earlier period to develop current costs, the rate of inflation must be accounted for by normalizing the historical cost to current equipment, material, and labor costs. This can be done using cost indexes for construction, materials, and labor (construction cost index [CCI]) published monthly by the *Engineering News-Record* (ENR 2014) as follows:

$$C_t = C_0(I_t/I_0) \quad \text{(Eq 6-1)}$$

where

C_t = estimated cost at present time t
C_0 = cost at previous time, t_0
I_t = Index at time t
I_0 = Index at time t_0

METHODS FOR ESTIMATING CLASS 4 AND CLASS 5 CONSTRUCTION COSTS

The scope of this chapter is estimating costs for the purposes of concept screening or feasibility studies, i.e., Class 4 or Class 5 level costs. The following three methods are appropriate for estimating Class 4 and Class 5 construction costs:

- estimation by cost–capacity equation based on previous projects,
- estimation by cost curves developed from historical data, and
- estimation by vendor quotes.

Depending on the information available, more than one of the methods may be used to develop a construction cost estimate. Class 1, 2, and 3 estimates are best developed using detailed unit costs with detailed takeoffs of all equipment and materials. Development of Class 1, 2, and 3 costs is beyond the scope of this manual.

Estimation by Cost–Capacity Curves Based on Previous Projects

The quickest and simplest method is to estimate cost based on a previous project using a cost–capacity curve. Many cost–capacity relationships plot as a straight line on log-log paper. Hence, the common form of the cost–capacity equation is:

$$C_2 = C_1(Q_2/Q_1)^x \quad \text{(Eq 6-2)}$$

where
C_1 = Cost at capacity Q_1
C_2 = Cost at capacity Q_2
X = Exponent

In the absence of better information, a value of 0.6 is often used for the exponent according to the rule of six-tenths. For modular unit processes such as RO, an exponent of 0.8 is sometimes applied.

The advantage of the cost–capacity curve method is that it is fast and easy. The disadvantage is that accuracy depends on similarity between the current and previous projects. The accuracy of this method increases with the level of knowledge about the comparison project. The cost–capacity curve method can be used to provide a reality check for a cost estimate developed by another method.

Estimation by Parametric Cost Curves

A more detailed cost estimate can be developed using parametric cost curves such as those given in McGivney and Kawamura (2008). These cost curves were developed by fitting an equation to historical cost data for a specific piece of equipment or treatment process. Each cost curve is fixed in time and must be correlated to the current time using ENR cost indices as specified in Eq 6-1.

The following example is a cost curve used to calculate the construction cost for sodium hydroxide (NaOH) feed:

$$\text{Cost in \$} = 118.16\,x + 38{,}701 \quad \text{(Eq 6-3)}$$

where
x = Dry NaOH feed in gallons per day

Estimation by Vendor Bids

A Class 4 or Class 5 cost estimate can be developed using vendor bids. In this method, a vendor bid is obtained for each major piece of equipment. After aggregating the cost of all equipment as supplied, costs must be added in to account for costs to develop the site and install and commission the equipment. Typically, these cost add-ons are estimated as a percentage of the total equipment cost. Table 6-3 shows examples of cost add-ons.

Table 6-4 shows an example of a construction cost estimate using this method. This example is for a brackish water RO plant with a total capacity of 10 mgd and consisting of five 2 mgd trains, with 85 percent recovery and deep-well injection of concentrate.

CONSTRUCTION COSTS FOR MAJOR PROCESSES USED IN DESALINATION AND CONCENTRATE MANAGEMENT

Some typical ranges for costs of major treatment processes that might be used in desalination and concentrate management are provided in this section. These costs are not meant to replace development of site-specific cost estimates by the methods discussed previously in this chapter.

Table 6-3 Examples of cost add-ons

Item	Category Description	Add-on Percentage Range (%)
1	Equipment installation[1]	35–65
2	Electrical power supply and instrumentation[2]	18–25
3	Site work[3]	2–10
Subtotal – Total direct cost[4]		
4	Contractor's general conditions[5]	2–10
5	Contractor's overhead and profit[6]	5–20
6	Bid market allowance[7]	0–5
7	Sales tax[8]	Per state
8	Contingency[9]	10–30
Subtotal – Construction cost[10]		
9	Engineering, administrative, legal costs (including construction management)[11]	25–35
Total project cost[12]		

[1] As a percentage of the total equipment cost. Percentage varies with complexity of the equipment. Generally lower percentages for skid-mounted equipment.
[2] As a percentage of the total equipment cost. Percentage varies depending on how much instrumentation is provided with the equipment and total power demand of the equipment.
[3] As a percentage of the total equipment cost. Percentage varies with condition of the site, amount and size of pipework, and distance to other facilities.
[4] Total direct cost is the sum of all equipment costs plus items 1, 2, and 3.
[5] General conditions are applied to the total direct cost. These include the cost of contractor mobilization, bonds and insurance, temporary onsite facilities, supervisory personnel, testing and startup, and other items included in the project specifications.
[6] Overhead and profit is the markup the contractor applies to the project and is a percentage of the total direct cost. The range typically depends on the bidding climate.
[7] Bid market allowance can be applied in times of a very active construction market to account for potentially higher costs.
[8] Sales tax is applied to the equipment costs based on the requirements of each state.
[9] Contingency allowance is applied to the total direct cost and varies depending on the status of the design process: higher contingency at planning level and lower contingency as the design team develops all areas of the project and the risk of unknown factors decreases.
[10] Total construction cost includes the total direct cost and items 4 to 8, as applicable.
[11] Engineering, administrative, and legal costs (including construction management) include the planning, design, and construction oversight costs as well as the owner's administration and legal costs and may include permitting.
[12] The total project cost is the sum of all application costs including item 9.

Microfiltration/Ultrafiltration Membranes

Table 6-5 shows typical construction costs for microfiltration/ultrafiltration (MF/UF) systems. The costs are based on hollow fiber or capillary membranes, and the cost includes installation and the membrane manufacturer's field services. The system includes the feed pump, strainer, valve skid, instruments, interconnecting manifold, backwash or backflush system and pipework, and the MF/UF rack with the membrane modules. Other equipment or systems to be integrated with the MF/UF equipment are pretreatment chemical dosing and membrane-cleaning systems. This equipment must be estimated separately and added to the cost of the MF/UF equipment.

Nanofiltration/Reverse Osmosis Membranes

Table 6-6 shows typical construction costs for NF/RO systems for desalinating brackish water. These installed costs include the cartridge filter, RO high-pressure pump, NF/RO membrane rack, pressure vessels, NF/RO elements, interconnecting high-pressure and

Table 6-4 Cost estimation example

Process	Applicable Range	Quantity per Unit Process	Number of Process Units/ Duty or Capacity	Process Cost per Unit*	Total Process Cost*
Raw water pumping	1–12 mgd	1	1 x 12 mgd	$250,000	$250,000
Chlorination	200–10,000 lb/ day	1	1 x 500 lb/day raw water 1 x 83.4 lb/day product water	$180,000	$180,000
Dechlorination	11–5,300 gpd	1	1 x 100 gpd	$25,000	$25,000
Acid dosing	11–5,300 gpd	1	1 x 120 gpd	$25,000	$25,000
Antiscalant dosing	11–5,300 gpd	1	1 x 120 gpd	$25,000	$25,000
RO trains	2 mgd trains	1	5 x 2 mgd	$1,800,000	$9,000,000
RO cleaning	1 train cleaning	1	1	$120,000	$120,000
Postchlorination	200–10,000 lb/ day			Use same facility as prechlorination	
Clear water storage tank (below ground)	0.01–8 MG	1	1 x 5 MG	$3,300,000	$3,300,000
Product water pumping	1.45–10 mgd	1	1 x 10 mgd	$200,000	$200,000
Brine disposal deep-well injection	Based on similar-sized deep-well injection wells	1	1,225 gpm	$3,600,000	$3,600,000
Administration, laboratory, and maintenance building	1–10 mgd	1	1 x 10 mgd	$200,000	$200,000

ENR CCI—Engineering News Record construction cost index
* In 2016 US$

Subtotal process cost (based on ENR CCI = 8,889):	$17,120,000
Subtotal process cost (adjusted to ENR CCI = 10,981):	$21,149,000
Pipeline work, 10%:	$ 2,115,000
Site work – landscaping, 5%:	$ 1,057,000
Site electrical and controls, 20%:	$ 4,223,000
Total construction cost:	$28,544,000
Engineering, legal, and administrative costs, 35%:	$ 9,990,000
Total project cost:	$38,534,000

Table 6-5 Typical MF/UF construction cost*

Capacity (mgd)	MF/UF Train Cost	Unit Cost ($/gpd)
0.5	$400,000	0.8
1.0	$700,000	0.7
2.0	$1,200,000	0.6
3.0	$1,500,000	0.5
5.0	$2,000,000	0.4
7.5	$2,850,000	0.38
10.0	$3,700,000	0.36

ENR CCI—Engineering News Record construction cost index, MF/UF—microfiltration/ultrafiltration
* 2014 ENR CCI = 10,732

Table 6-6 Typical brackish water NF/RO membrane construction cost*

Capacity (mgd)	NF/RO Train Cost	Unit Cost ($/gpd)
0.50	$600,000	1.20
0.75	$860,000	1.15
1.0	$1,100,000	1.10
1.5	$1,500,000	1.00
2.0	$1,800,000	0.90
3.0	$2,400,000	0.80
5.0	$3,500,000	0.70
7.5	$4,950,000	0.66
10.0	$6,250,000	0.65

ENR CCI—Engineering News Record construction cost index, NF/RO—nanofiltration/reverse osmosis
* 2014 ENR CCI = 10,732

Table 6-7 Typical seawater NF/RO membrane construction cost*

Capacity (mgd)	NF/RO Train Cost	Unit Cost ($/gpd)
0.25	$1,050,000	4.20
0.50	$1,750,000	3.50
1.0	$3,000,000	3.00
1.5	$3,750,000	2.50
2.0	$4,800,000	2.48
3.0	$6,900,000	2.46
5.0	$11,100,000	2.44
7.5	$15,800,000	2.42
10.0	$20,300,000	2.40

ENR CCI—Engineering News Record construction cost index
* 2014 ENR CCI = 10,732

low-pressure manifolds and pipework, control valves, isolation valves, instruments, and local control.

Table 6-7 shows typical construction costs for NF/RO systems for desalinating seawater. These costs are shown because RO can be used to desalinate RO concentrate in some concentrate-minimization strategies. In such cases, operating conditions and equipment requirements are more analogous to desalination of seawater than of brackish water. These installed costs include the cartridge filter, RO high-pressure pump, NF/RO membrane rack, pressure vessels, NF/RO elements, interconnecting high-pressure and low-pressure manifolds and pipework, control valves, isolation valves, instruments, and local control. These costs are not meant to represent the cost for seawater desalination; rather, they are shown to capture the cost increase because of higher operating pressures.

The costs in Table 6-6 are based on brackish water with TDS of 2,000–5,000 mg/L, with recovery of 80–85 percent and average membrane flux of 12.5 gfd. The costs include installation and the RO equipment manufacturer's field services.

The costs in Table 6-7 are based on feedwater with TDS of up to 35,000 mg/L with recovery of 40–50 percent, and average membrane flux of 8 gfd. The costs include installation and the RO equipment manufacturer's field services.

Solids-contact Softener

Chemical softening can be used to enhance RO recovery, either recovery of the primary desalination step or recovery of a secondary RO being used to treat concentrate from the primary RO. The solids-contact chemical softener used for this purpose is the same process used for softening freshwater. The solids-contact chemical softener or reactor clarifier consists of a flash mixing chamber with rapid mixer; a flocculation or reactor chamber where the sludge mixes with the incoming water stream to promote growth of flocs; a clarifier section, where the softened and clarified water is separated from the sludge; and a sludge thickener, where the sludge is scraped toward the bottom of the clarifier for removal.

Table 6-8 shows typical equipment costs for solids-contact softening systems. The costs do not include the chemical preparation and dosing system or civil works.

Pellet Softener

Table 6-9 shows equipment costs for a pellet softener. A pellet softener can be used to enhance RO recovery in lieu of solids-contact softening. The costs include the reactor, recirculation pump, valves, and pipework and fittings, and the cost information was derived from bid results for various projects. Chemical dosing, sludge dewatering systems, and installation are not included in the costs. Capacity depends on the level of hardness in the feedwater and the treatment goal.

Evaporation Ponds

Design factors and a preliminary level cost model for evaporation ponds were developed by Mickley & Associates (Mickley 2006). The cost model was developed for preliminary estimates only and intended to provide insight into the cost factors to support development of more detailed costs. A calculation worksheet was provided with associated graphs

Table 6-8 Equipment costs for reactor clarifier for installation in 15-ft deep concrete or steel wall*

Clarifier Diameter, ft	Capacity, gpm	For Concrete Tank Installation†	With Steel Tank
20	220	$115,000	$165,000
35	700	$200,000	$280,000
50	1,500	$290,000	$410,000
75	3,500	$440,000	$620,000

ENR CCI—Engineering News Record construction cost index
* 2015 ENR CCI = 10,981
† The costs do not include the concrete tank and the concrete base and foundation for the clarifier.

Table 6-9 Pellet softener equipment costs* including sand wash/feed pump and control

Diameter, ft	Price	Capacity, gpm
4	$650,000	300–500
6	$875,000	675–1,125
8	$1,150,000	1,200–2,000

ENR CCI—Engineering News Record construction cost index
* 2013 ENR CCI = 9,552

Table 6-10 Deep-well injection costs from recent installations

	El Paso, Tex.	San Antonio, Tex.	East Cherry Creek, Colo.	Vero Beach, Fla.	Sterling, Colo.	North Miami Beach, Fla.
Number of wells	3	2	1	1	2	1
Injection depth (ft)	3,700–4,000	4,200–4,800	10,500	1,650 and 3,000	6,000–7,000	2,858
Injection tubing ID (in.)	—	—	7	16.6	7	14.46
Injection flow (gpm)			200	840	200	
Cost (million US$)	$7.75 (3 wells)[*]	$4.4 (1 well)[†]	$3.2[‡]	$4.44[§]	4.5 (2 wells)[‖]	4.9[#]
Year	2007	2012	2011	2010	2011	2009

ID—inner diameter

* Drilling and well casing only; surface injection facilities, including tanks, controls, and piping, were an additional $4.94 million; injection tubing, instrumentation, and controls were an additional $1.55 million; all costs updated to 2012 values.

† For construction of the first well; an additional $0.64 million for planning, design, and permitting

‡ Does not include permitting, engineering, and construction management

§ Injection wells (drilling, casing) cost was $4.44 million; pipeline (about 3 miles) was $2.83 million; pump stations and emergency generators, 3 MG storage facility, $3.37 million; design, permitting, and construction services, $0.92 million

‖ Two wells drilled to provide redundancy

\# Capital expenditure (minus pump and pipe) was $4.9 million; piping and pump costs ($350,000) were minor as the system is in close proximity to the desalination plant.

for estimating costs. In addition, a closed-form regression equation based on use of the worksheet over a considerable range of cost-factor values was given. Please see this document for evaporation pond costs.

Deep-Well Injection

Table 6-10 shows recent deep-well injection construction costs from six sites. The data come from case studies published in the WateReuse report *Desalination Concentrate Management Policy Analysis for the Arid West* (2015).

Mechanical Vapor Compression Brine Concentrators and Crystallizers

Equipment costs for mechanical vapor compression (MVC) brine concentrators and crystallizers obtained over the years from vendors are shown in Tables 6-11 and 6-12. These were informal quotes for study purposes rather than bid quotes. The costs do not include installation.

ANNUAL COSTS

Annual costs that must be added to the amortized construction cost to derive the total treatment cost include debt service and O&M costs.

Debt Service Cost

The debt service cost is the amount paid for the capital and interest for financing the project, which is based on the total project cost. Typically, water plants are financed for

Table 6-11 MVC brine concentrator equipment costs

Capacity, gpm	Equipment Cost, $M	Unit Cost, $/gpd	Year
705	8.25	8.13	2002
433	6.66	10.65	2002
345	5.84	11.75	2002
191	5.15	18.72	2002
750	11.4	10.55	2010
1,500	17.3	8.01	2010
716	9.4	9.08	2014

MVC—mechanical vapor compression

Table 6-12 MVC crystallizer equipment costs

Capacity, gpm	Equipment Cost, $M	Unit Cost, $/gpd	Year
35	2.4	47.62	2002
39	3.0	53.42	2002
35	4.4	87.30	2010
287	12.5	30.30	2014

MVC—mechanical vapor compression

30 years and the prevailing interest rates range from 3 to 10 percent. The monthly payment is calculated using the following formula:

$$A = P*\{[r(1+r)n]/[(1+r)n-1]\} \quad \text{(Eq 6-4)}$$

where

A = monthly payment
P = amount financed
r = interest rate
n = total number of payments

Operation and Maintenance Cost

Operation and maintenance cost includes direct and indirect costs. Direct costs are the chemicals and consumables, membranes, spare parts, and power costs. Indirect costs include manpower and other costs associated with operating the plant.

Chemicals and consumables. The chemicals used in the treatment plant include those continuously added in the various process steps, chemicals used intermittently such as cleaning chemicals and laboratory reagents, and general maintenance chemicals. Consumables are materials that are regularly replaced; they are either consumed or at the end of their useful life (e.g., cartridge filter elements). Table 6-13 shows an example of a cost estimate for chemicals.

Membrane replacement. The life of an RO membrane is typically five to seven years. Therefore, a plant can expect to replace 14–20 percent of its membranes each year. Warranties from membrane suppliers are often prorated over the life of the warranty. For example, if a membrane warrantied for five years must be replaced after four years, the value of the warranty is 20 percent of the cost of the membrane.

Table 6-13 Chemical cost example based on 10 mgd brackish water RO plant

Chemical & Consumables	Dosage	Quantity	Unit Cost*	Annual Cost*
Chlorine gas, raw water	3 mg/L	294 lb/day	$0.395/lb	$42,422.05
Sulfuric acid, 98%	15 mg/L	1,471 lb/day	$0.08/lb	$42,958.46
Sodium metabisulfite, 40%	20 mg/L	195 gal/day	$1.91/gal	$136,206.03
RO anti-scalant	2 mg/L	294 lb/day	$1.20/lb	$128,877.12
Chlorine gas, treated water	1.5 mg/L	125 lb/day	$0.395/lb	$18,036.29
Cleaning chemicals				$50,000.00
Oil & grease				$2,000.00
Cartridge filter elements		1,700/yr	$7.50	$12,750.00
			Total cost	$433,249.95
				$0.12/1,000 gallons

RO—reverse osmosis
* 2016 US$

Table 6-14 Manpower cost example based on 10 mgd brackish water RO plant

	Number	Total Annual Cost*
Plant manager	1	$146,000.00
Plant operators	4	$408,800.00
Electrician	1	$87,600.00
Mechanic	1	$87,600.00
Laborer	3	$157,680.00
Laboratory technician	1	$73,000.00
Total	10	$960,680.00
Plus benefits (50% of salary)		$480,340.00
	Total annual labor cost	$1,441,020.00
		$0.79/1,000 gallons

RO—reverse osmosis
* 2016 US$

Labor cost. Labor cost is highly variable from site to site depending on local pay scales, whether employees are unionized, and the number of employees required to operate the plant. Staffing requirements depend on the level of plant complexity and automation and whether the plant is operated full-time or part-time. Table 6-14 shows an example of an estimate of manpower cost.

Power cost. Power cost can be estimated by summing the major electrical loads through the plant. These might include RO high-pressure pumps, chemical pumps, product water delivery pumps, heating, ventilating, and air conditioning, and lights. Often the RO high service pump contributes a significant portion of the power cost. RO membrane manufacturers have computer software that can be used to model the power consumption of the RO high service pumps. Table 6-15 shows an example of a power consumption estimate for chemicals.

Table 6-15 Power consumption example based on 10 mgd brackish water RO plant

Drive	Number of Equipment/Hours Operating per Day	Power Rating	Power Consumption kWh/d
Raw water pumps	5/24 hr	75 hp	5,059.2
RO high-pressure pumps	5/24 hr	300 hp	23,716.8
Chlorine booster pump raw water	1/24 hr	7.5 hp	123.8
Acid dosing pump	1/24 hr	0.3 hp	4.8
Anti-scalant dosing pump	1/24 hr	0.3 hp	4.8
SMBS dosing pump	1/24 hr	0.3 hp	4.8
Chlorine booster pump – product water	1/24 hr	5 hp	74.4
Flushing pump	Intermittent	50 hp	—
Brine discharge pump	1/24 hr	1,200 hp	19,183
Treated water pumps	5/24 hr	60 hp	4,395.6
RO cleaning pump	Intermittent	150 hp	—
RO cleaning tank heater	Intermittent	45 kW	—
Air compressor	1/8 hr	7.5 hp	40
Controls	1	25 kW	480
Others	1	35 kW	720
	Total power consumption	53,807.9 kWh/day	
		5.38 kWh/1,000 gallons	

RO—reverse osmosis

REFERENCES

AACE (Association for the Advancement of Cost Engineering). 2005. *Cost Estimate Classification System—as Applied in Engineering, Procurement, and Construction for the Process Industries*. www.costengineering.eu/Downloads/articles/AACE_CLASSIFICATION_SYSTEM.pdf (accessed Feb. 22, 2019).

ENR (Engineering News-Record). 2014. *Construction Economics Weekly*. http://enr.construction.com/economics/current_costs/ (accessed May 9, 2018).

McGivney, W.T. and Kawamura, S. 2008. *Cost Estimating Manual for Water Treatment Facilities*. Hoboken, N.J.: John Wiley & Sons, Inc.

USBR (US Bureau of Reclamation). 2006 (2nd ed.). Membrane Concentrate Disposal: Practices and Regulation. Desalination and Water Purification Research and Development Program Report. No. 123. Denver: USBR.

WateReuse. 2015. Desalination Concentrate Management Policy Analysis for the Arid West. Alexandria, Va.: WateReuse.

Chapter 7

Regulatory, Safety, Operational, and Environmental Issues

Dave Stewart, Fred Bloetscher, and Melanie Goetz

This chapter focuses on the regulatory requirements of permitting an inland brackish water treatment plant at both the federal and state levels, including deep-well injection processes. An overview of the environmental impact associated with brackish water treatment is included, along with the operational and safety issues associated with this process.

CURRENT CONCENTRATE MANAGEMENT ISSUES

The regulations associated with inland brackish water treatment focus mainly on the management of concentrate from the facility. There are numerous methods to manage this material, but each has its unique set of issues, depending on the federal, regional, and local regulations; chemistry of the concentrate; and how the eventual management will evolve.

The available literature associated with concentrate management and disposal is somewhat limited, which restricts the ability of researchers to better understand concentrate disposal options. Squire (2000) reports the disposal methods in the United Kingdom in Table 7-1.

In comparison, existing methods for brackish water concentrate disposal in the United States include deep-well injection, direct and indirect outfalls into surface water

Table 7-1 Disposal methods in the United Kingdom

Disposal Method	Percentage
Surface water	48
Wastewater treatment plant	23
Land application	13
Deep well	10
Evaporation ponds	6

bodies (primarily the ocean), reuse with a growing body of ZLD strategies for inland areas, and methods for salt recovery (refer to Chapters 3, 4, and 9).

For inland brackish water treatment, the most feasible and economic choices are limited to controlling the concentrate. These choices include deep-well injection, wastewater treatment plant industrial pretreatment, evaporation ponds, and use of the brine for mineral recovery and reuse.

REGULATORY ISSUES

There are several federal, state, and local agencies that will need to provide approval or permitting of any surface discharge. Federal and state agencies include at a minimum the US Fish and Wildlife Service, the US Army Corps of Engineers, the USEPA, and state environmental agencies. The permits required are often project- and site-specific. Permitting regulations in other countries are limited.

The lack of available options for disposal of concentrate from desalination plants has limited the pursuit of RO of waters with impaired or brackish quality as a water supply in water-stressed areas. The limitation with disposal of concentrate results from its designation as an industrial waste; the acute toxicity of the concentrate, which limits discharge to receiving waters; the lack of access to disposal options, such as outfalls and deep injection wells that can mitigate certain toxicity issues; and the construction costs of treatment systems.

Dilution can help where current outfalls exist for brackish water treatment near the ocean, but the toxicity remains an issue. To mitigate potential toxicity issues, discharge of concentrated seawater to the ocean will likely require a dispersion field to ensure minimal variation from background water quality at the point of discharge. Factors to consider when determining the location of the dispersion field include fishing, shrimping, and mollusk harvesting as well as recreational and commercial boating activities of the community. The dispersion field must be located to allow periodic inspection of the diffuser nozzles and to limit the impact of sand and ocean debris covering the diffusers.

Specific rules for concentrate disposal into surface waters are absent in many states because there has been no need for them to date. Where the issue of concentrate disposal arises, antidegradation rules are often used to define discharge limitations.

Often, the salt content of concentrate precludes discharge to any surface water bodies, necessitating other options. Deep wells can be an option if the geology will allow for injection of the concentrate into the subsurface. Forty states have some degree of primacy from the USEPA for injection wells. Concentrate is normally discharged in Class I injection wells (the advantages and disadvantages of deep-well injection are discussed in Chapter 4). In all of these cases, the site restrictions of proximity to the ocean and limitations of geology may preclude one or both disposal methods. Readers are encouraged to have conversations with regulatory personnel before proceeding with a project to determine the needs for permitting any concentrate options.

For other inland brackish water systems, the use of the Class I injection well, evaporation systems (both natural and enhanced), and use of the concentrate for chemistry are options explored in later sections of this chapter.

FEDERAL REGULATIONS

The regulatory issues associated with inland desalination and concentrate management fall mainly under the CWA. The concentrate will be controlled as an industrial waste as this waste has not been identified specifically as a byproduct according to the CWA. Some states, such as Florida and California, specifically control industrial waste. However, an amendment to the CWA would be required for industrial waste to be specifically identified under the CWA.

Concentrate management falls into the following areas of regulatory management.

1. CWA. If the concentrate is to be discharged to a surface water, an NPDES permit is required. If the concentrate is to be discharged to a sewer system, then the user is required to obtain an Industrial Pretreatment Permit as a significant industrial user (SIU).
2. SDWA. The SDWA controls the Underground Injection Control (UIC) well system. Concentrate management would likely use either a Class I or Class V injection well, depending on the constituents being disposed of in the well.
3. Resource Conservation and Recovery Act (RCRA). Although concentrate waste streams are not normally considered a hazardous waste, they are required to be classified when disposed of as a waste stream. The utility is required to ensure that the wastes are being properly handled. Solid waste disposal is defined within the RCRA regulations.
4. Evaporation ponds. Evaporation ponds have been effective in the disposal of concentrate for inland desalination facilities. These disposal ponds have specific requirements as they relate to leak detection and the potential for air emissions from the concentrate.
5. Comprehensive Environmental Response, Compensation, and Liability Act (CERCLA). CERCLA is generally referred to as "Superfund." Normally, a concentrate stream would not be classified under CERCLA. However, when the constituents of concern are high enough in concentration to be classified as a "release" under CERCLA, the utility must verify that it does not have this type of material in the concentrate stream.
6. Technologically enhanced naturally occurring radioactive material (TENORM), as radioactive particles such as those of uranium can exist in the source water. These particles would be rejected by the membrane system and concentrated in the discharge stream. The utility is required to investigate the quantity of TENORM in the concentrate to ensure that the Nuclear Regulatory Commission regulations are not being violated.

In addition to concentrate management, the utility will also be required to manage the other waste streams, such as the cleaning and conditioning agents, as well as particulate matter that is rejected by the membrane system. Each of these waste streams must be characterized per 40 Code of Federal Regulations (CFR) Part 261 (RCRA characterization testing and listed wastes). The various federal rules and regulations per the USEPA are outlined in the following sections. Most states have adopted similar rules and regulations as well.

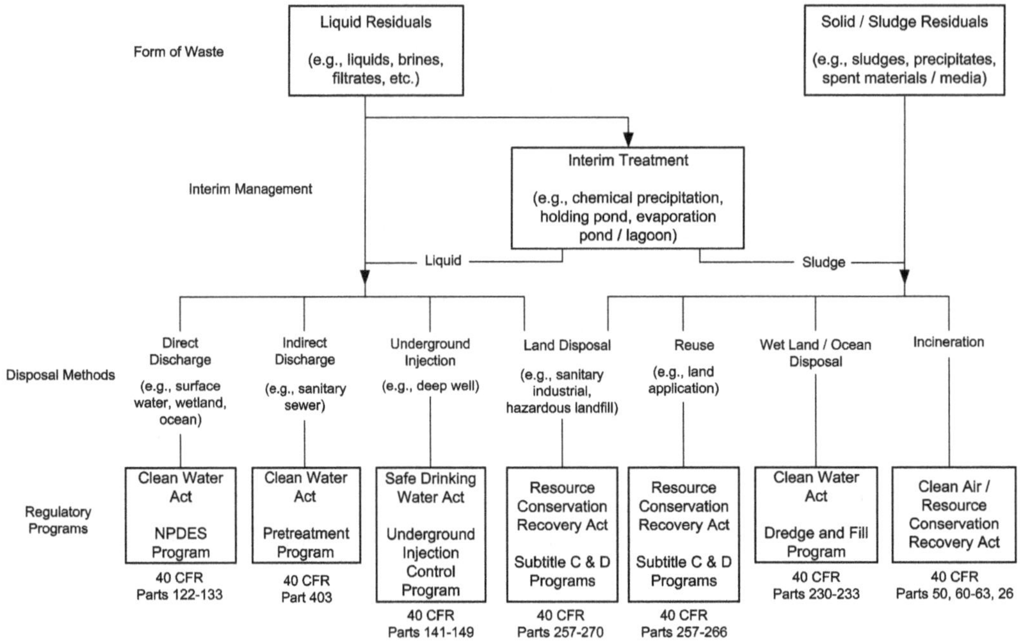

Figure 7-1 USEPA classification of waste streams

The pretreatment system prior to the membrane system can also generate solids. All of these solids need to be classified per the USEPA and state/local regulations.

Form of Waste

The first issue that the utility will face is determining the form of the waste. The waste can be in the form of liquids, such as concentrates, filtrates, etc. The utility will also be required to manage other waste streams generated by the desalination plant operations such as the spent membrane cleaning and conditioning chemicals, as well as sludges, spent materials, precipitants, and other solid media. The EPA will require that all types of waste are identified by the various regulations as shown in Figure 7-1 and detailed as follows.

Clean Water Act

The CWA will cover the discharges to surface water or to a POTW. For direct discharge, an NPDES permit will be required. This will fall under 40 CFR Parts 122 to 133. Typically, the state of origin will have this permit system and will issue a state permit that is allowed under the USEPA federal permit system.

Two types of NPDES permits can be obtained: individual and statewide. The individual permit, which allows use of the stream flows and potential mixing zone of the receiving water, can take as long as 18–24 months to obtain. Statewide permits are uncommon but require specific discharge parameters that meet the permit limits at the discharge point. Receiving water mixing zones are not considered in this type of permit.

If the utility discharges this water to a POTW, then an industrial pretreatment permit will likely be required. The EPA industrial pretreatment program is controlled under 40 CFR Part 403. While this user will not be classified as a categorical discharger, it will likely qualify as an SIU. In some cases, verification that the concentrate or solids do not "pass through" the POTW to a receiving stream—which has been found to occur with TENORM material—will be needed. In the case of conventional pollutants, because of the RO concentration of the brine, there are potential priority pollutants that can affect a discharge to a POTW. Therefore, the permit applicant will need to determine whether any of the 126 priority pollutants found in 40 CFR Part 423 exist in the concentrate stream and, if so, at what concentration. Typically, these pollutants are also found as compounds excluded from discharge because of the "pass through" issue. In addition, for inland desalination projects, nitrates are typically found in the source water and can affect the treatment process at the POTW.

WET testing is also part of the permitting process; it is detailed in 40 CFR 136.3. This test will subject *Ceriodaphnia* and fathead minnows to the concentrate stream. If the discharge is to a saline water body, the test will be performed with mysid shrimp. The WET tests have both an acute and chronic component to determine whether the concentrate is toxic to the aquatic life in the stream and, if so, how toxic. If the concentrate has an LC50 above a certain tolerance, a means to identify and then reduce the toxicity must be employed. This process can be time consuming but likely will be required for any stream discharge permit.

Safe Drinking Water Act

The SDWA controls the UIC well permit system. There are six types of wells under the UIC, but concentrate management typically uses either a Class I or Class V well. These wells are defined as follows.

- Class I: Inject hazardous wastes, industrial nonhazardous liquids, or municipal wastewater beneath the lowermost underground source of drinking water (USDW). The vast majority of injection wells for concentrate management will be Class I wells.
- Class V: All injection wells not included in Classes I–IV. In general, Class V wells inject nonhazardous fluids into or above USDWs and are typically shallow, onsite disposal systems. However, there are some deep Class V wells that inject below USDWs.

USDW is generally defined as a water that has TDS of less than 10,000 mg/L. It is assumed that a groundwater source with this concentration of TDS can be used as a water resource.

UIC well permits can be difficult to obtain and require a significant amount of time in the permitting process; protection of the drinking water aquifer is sometimes difficult to demonstrate. Monitoring wells are usually required around the Class I or V injection wells to ensure that the drinking water aquifer is not being contaminated. The utility has to ensure that the injection pressure will not create fissures that will contaminate other layers near the injection well. An excellent example of this type of injection well for concentrate disposal for inland facilities is found at the Kay Bailey Hutchison Desalination Plant in El Paso, Tex. (see case study in Chapter 8).

Resource Conservation and Recovery Act

In the federal regulations, RCRA controls solid wastes. However, liquids can also be defined as solid waste; therefore, concentrates can be regulated under RCRA. Solid wastes typically will be disposed of at a landfill. These landfills are regulated under 40 CFR Parts 257 to 270. While liquid streams can be disposed of at a landfill, this is normally discouraged as they can cause issues with landfill leachate. However, in some cases in the pretreatment of brackish waters, where a precipitation step for the removal of scale-forming compounds is implemented, the solids from this precipitation step will be disposed of as a nonhazardous solid waste and controlled under RCRA.

As part of this process, the waste will need to be classified under 40 CFR 261.1, where testing for the characteristics of a hazardous waste is found. If the waste is found to be hazardous, then it must be disposed of at an RCRA-permitted facility. This classification will increase the cost of disposal significantly, sometimes up to ten times the cost of just a solid waste disposal fee. This can also occur because of the concentration factors of the concentrate solution. Therefore, it is important to characterize the source water and to ensure that the concentrates and solids generated by the process do not create a hazardous waste.

Land application of the concentrates is also potentially available to the utility if the chlorides are low. However, this method is rarely used and likely would not be allowed because of the concentrations of the constituents.

Most concentrate solutions are high in sodium content. The goal for the future will be to use these concentrate solutions to form a byproduct that can be used in other industries. These alternatives are discussed in detail in Chapter 9. An example of another potential process for the handling of this material, known as the Solvay process (Ahmed et al. 2001), uses ammonia and carbon dioxide to form sodium carbonate. This technology is worthy of further exploration.

Another technique for the disposal of concentrate is the use of evaporation ponds. In the western United States, the net evaporation of water typically can be more than 20 in. per year. This net evaporation can be enhanced through the use of mechanical dispersal systems. Using this natural evaporation process can be a cost-effective method for the disposal and concentration of the concentrate stream. The solution will reach a concentration of 330,000 mg/L of TDS and then will form a crystal salt solution that can be stabilized with mixtures of fly ash. This concentrate stream can also be used in the oil and gas industry as a 10-lb brine (220,000 mg/L TDS concentration) for drilling operations or can be made into sodium hypochlorite and sodium hydroxide through the chlor-alkali process. These byproducts have significant value in the marketplace. This technology is reviewed and discussed in Chapter 9.

Another enhanced evaporation system is the Altela system (Finley 2015). This system was piloted at the East Cherry Creek Valley Water & Sanitation District (ECCV) RO facility in Colorado. The concentrate disposal system was a Class I injection well. However, the Altela system allows for additional water recovery while reducing the amount of concentrate disposal (Finley 2015).

When providing a byproduct from the concentrate stream, the Altela system will remove the material from the USEPA regulations for waste disposal (RCRA). This is because of the recycling of material, which is specifically excluded from 40 CFR Part 261.

Comprehensive Environmental Response, Compensation, and Liability Act

CERCLA is typically applicable for the remediation of a site or facility; it requires the cleanup of materials that can harm the environment. This requirement will only affect a utility if the concentrate stream, or one of the other associated waste streams, is released in an uncontrolled manner into the environment. While this is unlikely, the utility should be aware of the implications of this type of requirement in the planning of a waste management program.

Technologically Enhanced Naturally Occurring Radioactive Materials

TENORM can be regulated under the Clean Air Act (CAA), CWA, SDWA, or CERCLA. Naturally occurring radionuclides, such as uranium, radium 226, or radium 228, can be concentrated through the membrane process. This concentration can then require special handling of the concentrate waste stream. It is critical for the utility to test the source water and then determine the concentrate stream to ensure that the TENORM standards are not being exceeded.

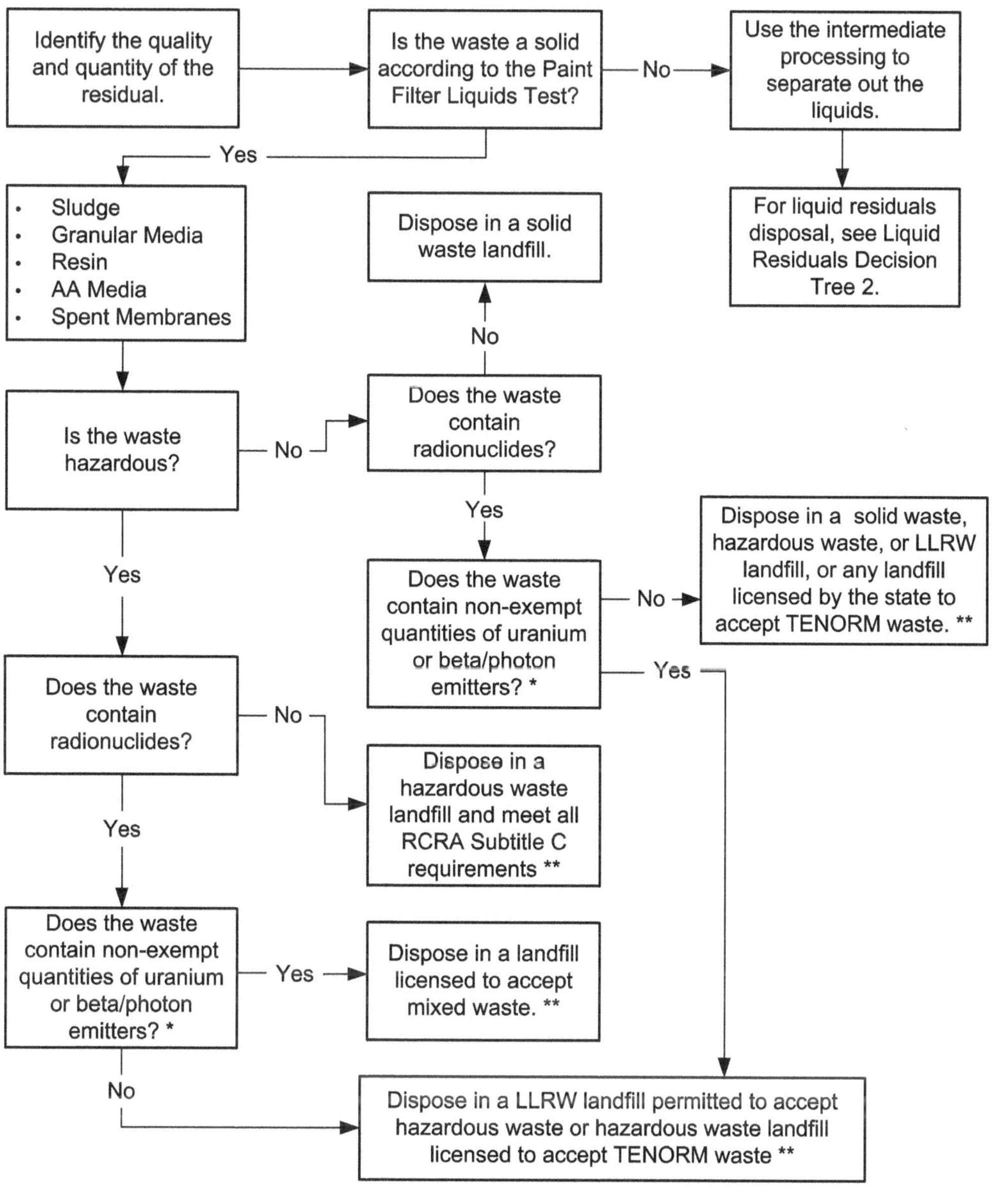

* Check with the state Radiation Program to see if beta/photon emitters are considered byproduct material and advise system to contact the NRC Regional Office or relevant Agreement State Agency to discuss potential licensing requirements.

** LDR Treatment Standards also apply. Check with the state Radiation Program to determine the proper disposal methods for waste containing radionuclides and hazardous waste.

Figure 7-2A Solid residuals TENORM decision tree

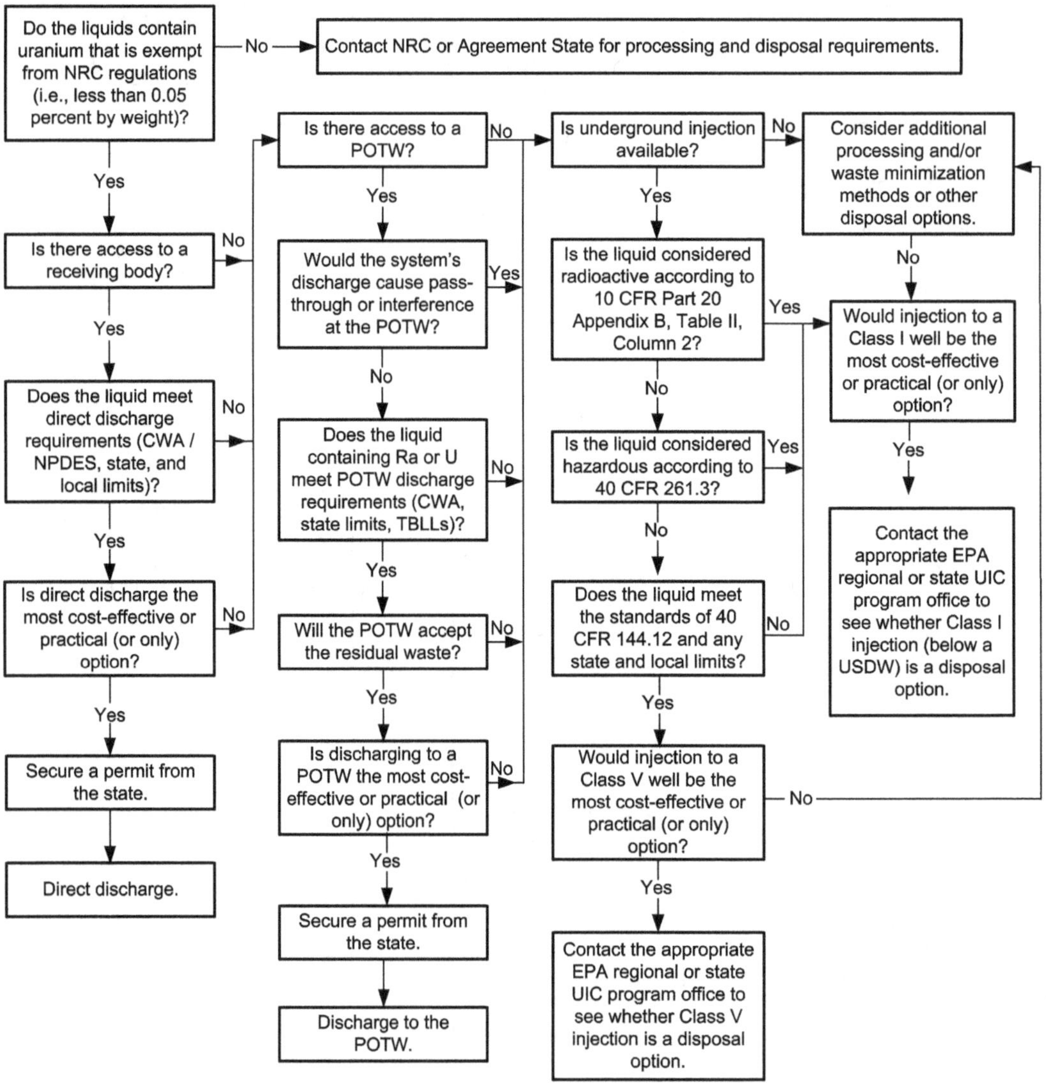

Figure 7-2B Liquid residuals TENORM decision tree

The processes of identifying the impact of TENORM are provided in Figures 7-2A and 7-2B. As noted in these figures, the permit holder will be responsible for the water stream but will also need to investigate the impact on the used membrane disposal, disposal of ion-exchange resins, sludge disposal, and any other identified waste streams.

STATE REGULATIONS

The state regulations vary across the United States. Most states follow the federal regulations, but in some cases, the state may be more restrictive. California is a good example of a state that has regulations that exceed the requirements of the federal regulations. Utilities are encouraged to discuss the issues with state agencies before pursuing these options. In many cases, the permitting of solutions will be a cooperative process as the concept of permitting concentrate is outside the current regulations.

Texas Regulations for Residuals

The Texas Commission on Environmental Quality (TCEQ) is the regulatory authority for the handling of concentrate for disposal. In most cases, this applies to evaporation ponds. The oil and gas industry also uses evaporation ponds for the handling of concentrates (both disposal and manufacturing) and is regulated by the Railroad Commission of Texas (RRC). The waste handling systems are required to be designed by a licensed professional engineer. These regulations also apply to the manufacturing and storage of products, such as ice-melting compounds for roads, agricultural fertilizer, and chemical manufacturing.

Class I and Class V injections wells as defined by the SDWA are regulated by the TCEQ. Class II injection wells for oil and gas operations are regulated by the RRC. This requirement is found under Chapter 27 of the Texas Water Code. Chapter 26 of the code regulates treatment facilities such as concentrate evaporation ponds.

These regulations identify specific requirements for the disposal of the concentrate. For example, the Class I injection well will require aquifer characteristics, such as how the concentrate will interact with the aquifer, and the casing and construction of the well will need to be provided to ensure that contamination of other aquifers will not occur.

The evaporation pond also requires that a licensed professional engineer provide the design and ensure that the pond will not contaminate surface or local unconfined groundwaters. Typically, the design will require a double liner system with an interstitial leak detection system. Texas requires that this liner system be as stringent as the Type I landfill managing Class I industrial solid waste (TCEQ 2007).

Colorado Regulations for Residuals

Colorado is similar to Texas in that the two disposal methods contemplated include the disposal of concentrate in a Class I injection well or an evaporation pond. Both types of wells are regulated by the Colorado Department of Public Health and Environment.

One issue identified in Colorado is the increase in earthquake activity because of saltwater injection wells from the oil and gas industry. The concentrate waters are lubricating the faults in the geologic formations, thereby causing earthquakes. This phenomenon of earthquake activity was first discovered in Colorado at the US Army Rocky Mountain Arsenal in the 1960s. The earthquake activity is usually associated with Class II injection wells for the oil and gas industry but should be a consideration for the Class I injection well as a potential issue.

Colorado also has allowed the concentrate to be mixed with natural water streams, resulting in a dilution of the concentrate stream. The ECCV water authority applied for and obtained a permit for blending the concentrate with the canal releases from the Barr Lake & Milton Reservoir system (DiNatale Water Consultants 2012). Because of the opposition of the agricultural community, the ECCV determined that a Class I injection well was a more viable solution.

California Regulations for Residuals

The California regulations for the disposal of residuals follow the other western states. However, there are numerous agencies involved in the permitting and disposal of concentrate. They include agencies such as the California State Water Resources Control Board (which controls the NPDES permit system), the California Coastal Commission, various regional water quality control boards, the California Department of Public Health, and various conservation commissions (California Department of Water Resources 2013).

As with Florida, there is the potential for inland as well as ocean disposal in California. As it relates to concentrate disposal, the water plan comprises the following options:

- discharge to separate permitted wastewater collection and treatment systems;
- discharge and dispersion to water bodies, such as oceans and bays;
- discharge by land application, usually involving further solids disposal after evaporation of liquid portion of discharge;
- discharge to Class I injection wells; and
- disposal by use of ZLD techniques, whereby solids are the only material disposed of from the process.

The water plan identifies that most of the concentrate disposal will be discharged to the municipal wastewater treatment system, where it is incorporated with other municipal wastewaters for treatment and discharge.

Florida Regulations for Residuals

The Florida Department of Environmental Protection (FDEP) regulates concentrate disposal as a "non-hazardous potable water byproduct." Municipal utilities are responsible for determining that they are not generating a hazardous waste (see RCRA and TENORM regulations). In most cases, the RO concentrate is disposed of through a UIC Class I injection well (FDEP – 62-528 FAC). In Florida, the Class I injection well can accept either domestic or industrial wastewater. The RO concentrate qualifies as an industrial waste. The injection well needs to be below the lowermost aquifer within ¼ mile of the well bore of a known drinking water source. A drinking water source is defined as an aquifer with less than 10,000 mg/L of TDS.

Most injection wells in Florida utilize the "Boulder Zone" as the preferred formation. This is typically found at depths near 2,500 ft below the ground surface. Vertical transport of injected water can occur through slow porous media flow. The utility will need to compare the buoyancy of the injected water to the formation water. Monitoring wells are normally required to ensure that the upward migration is not occurring.

Inland disposal of RO concentrate is possible in Florida but the disposal of this material will need to meet the criteria for both acute and chronic toxicity testing (602-302,300 FAR). In addition to toxicity testing, low levels of dissolved oxygen and high levels of hydrogen sulfide have prevented surface water discharges (Mickley et al. 1993). Enforcement of the water quality permit is not required at the point of discharge as Florida utilizes a mixing zone for the permit testing and compliance.

Regulations of Other Local Authorities

As mentioned previously, municipalities can also discharge the concentrate solution to the POTW. This discharge would be controlled under the industrial pretreatment regulations under 40 CFR 403. The concentrate stream would likely become an SIU classification. The SIU requires assurance that there is no potential for pass-through of the contaminants in the concentrate stream. A good example would be the TENORM parameters listed previously in this chapter. The POTW and the SIU would need to ensure that the NPDES permit is not violated as a result of the discharge of the concentrate to the municipal wastewater.

RELEVANT INTERNATIONAL REGULATIONS

International regulations also address concentrate management issues. In most cases, international management of the concentrate is similar to that of the United States. The

Middle East has been investigating the methods of concentrate management for several decades. Concentrate has been discharged to the Red Sea and is potentially causing environmental issues (Dawoud and Al Mulla 2012). Alternative concentrate management techniques are being investigated for reactions that produce byproducts, thereby eliminating concentrate disposal. Research is being conducted on the reaction of carbon dioxide in the presence of ammonia for the generation of sodium carbonate (i.e., the Solvay process mentioned previously).

Research in Sri Lanka has identified the same methodology as mentioned previously. Both Sri Lanka and Singapore are running out of drinking water, and concentrate salt solution management is a significant issue. In addition to the previously mentioned technologies, several unique solutions have arisen.

- The use of biological activated carbon in conjunction with capacitive deionization enabled removal of 90 percent of the concentrate salt solution along with its associated organics. This method allows for handling in other normal wastewater treatment processes.
- Membrane distillation is another technique being investigated internationally. Distillation will allow for very high concentrations of the concentrate stream.
- The chlor-alkali process is being investigated in France as a generator of a byproduct from the concentrate solution. This process will generate either hydrochloric acid or hypochlorous acid (bleach) along with sodium hydroxide. This may be useful for the water treatment plant, which will use the chlorine for disinfection purposes.

FUTURE PERMITTING CONSIDERATIONS

The future of concentrate and residuals management is a significant issue. It is projected that the use of inland desalination will be limited because of the issue of concentrate management. Therefore, topics addressed in Chapters 5, 6, and 9 will be important factors when considering concentrate management in the future. This is especially true for concentrate being used for byproduct manufacturing.

ENVIRONMENTAL IMPACTS, OPERATIONS, AND PERMITTING REQUIREMENTS

Nanofiltration/Reverse Osmosis, Ocean Outfalls, and Inland Surface Waters

Concentrate discharge considerations. Management of the concentrate produced by desalination plants can be a challenge because of the increased regulation of discharges and heightened public concern. A natural disposal site for the concentrate is the ocean. Ocean discharges have two primary differences from other surface water discharges: higher density of the receiving water (principally because of salinity) and much greater volume of receiving waters. The buoyancy of the plume, along with marine currents and turbulence, results in three distinct phases of dilution: jet, rise, and near field. Far-field movement generally has been ignored because of the inability to trace it. However, appropriate technology is required to provide proper dispersion of the concentrate so that the environmental effects are minimized. Regulatory agencies will require the appropriate modeling and operational protocols to demonstrate an environmentally acceptable discharge option.

Directed or collocated outfalls could potentially affect aquatic life, including benthic organisms. Disposal through collocated outfall encompasses discharging through an existing wastewater treatment plant or power plant outfall. Discharge through a wastewater plant is practiced in Dania Beach, Fla.; in Hollywood, Fla. (after blending with treated wastewater); and at the Tampa Bay Seawater Desalination Plant (after blending with power plant cooling water).

There are three barriers to disposal to surface water bodies, even in the ocean: toxicity (including ion toxicity), pH, and concentrate density. The pH issue is the result of acid addition to prevent scaling of calcium carbonate. Excessive TDS concentrations may have benthic impact because of density differentials. Ion imbalance is more acute with brackish water discharges.

Toxicity defined. The type and quality of the constituents in the raw water is the biggest factor to consider. The treatment that the source water undergoes before reaching the membrane also affects the "quality" of the concentrate. The first major problem is pH because concentrate from brackish desalination plants is relatively acidic when it leaves the membranes. Low pH from pretreatment processes can cause toxicity in aquatic environments. The next problem commonly faced when dealing with concentrate, especially generated from the desalination of brackish water, is that the concentrate is very condensed and may exceed the salinity of neighboring water bodies. High-density water from brackish processes may have TDS in excess of potential receiving waters, so it may have benthic impact. Even for fresh and slightly brackish water sources, a high level of TDS and salts is toxic to grass, crops, and landscaping and is regulated as such.

Imbalance in the ions contained in the concentrate discharge is another area of concern. Mickley (2000) and FDEP (2004) identified the likely cause as an ion imbalance in the concentrate because membranes remove the majority of the ions within the water; therefore, the ratio of various ions to the concentrate TDS is not in the same proportions as the receiving waters. Through the testing of ions' and concentrations' effects on aquatic life, Pillard et al. (2002) showed that calcium, fluoride, and potassium were all probable causes of toxicity. In the 2002 study, Pillard et al. studied a variety of sources (234 desalination facilities) for the various concentration differences and disposal techniques. However, there are differences between the water studied by Pillard et al. (2002) and potential RO concentrate from brackish sources. The AWWA study (2008) used nanofiltered freshwater as opposed to the brackish water used by Pillard et al. (2002) as the water source. For brackish water systems, the toxicity will include the added ions of sodium and chloride in addition to the aforementioned ions.

Pillard et al. (2002) showed that calcium and fluoride created toxicity as a result of ionic imbalance. The imbalance results from the fact that all creatures tend to have ionic balances that match that of seawater. An ion imbalance in the concentrate becomes an issue because membranes remove the majority of the ions within the water and therefore the concentration of ions is not in its correct proportions with the receiving water. When the concentrate has a different ion balance, an imbalance occurs.

In general, toxicity to aquatic organisms is the principal concern where RO concentrate is discharged to surface waters and ocean environments. Toxicity can arise from specific constituents or combinations of constituents in the RO concentrate (e.g., organic constituents targeted for removal, such as pesticides) or common ions such as ammonium, calcium, potassium, and arsenic. Toxicity from common ions, with the exception of ammonium, generally is not an issue because of the initial dilution provided by the receiving water body and high background concentrations in discharges to estuarine and marine water bodies (Mickley 2009). The extent of toxicity in commingled RO concentrate and receiving waters cannot be predicted in advance. Concentrate toxicity needs to be evaluated by laboratory testing with freshwater or marine aquatic test organisms appropriate for the receiving waters.

Estimates were made using Reverse Osmosis System Analysis (ROSA, The Dow Chemical Co., Midland, Mich.) to interpret what sort of treatment would be needed for the departing concentrate. The waters studies include: hard fresh Biscayne Aquifer water, which is similar to that used in Mickley (2003); recycled Biscayne Aquifer water; brackish

Floridan water with a TDS concentration of 2,400 mg/L; and seawater. The first two are readily collectable from the City of Hollywood, Fla., from their NF and low-pressure RO treatment skids. Table 7-2 outlines the water quality of each and the concentrate predicted from the ROSA software. Table 7-3 outlines the concentrate deviations from seawater, which is the method used by Pillard et al. (2002) to predict toxicity.

From these data, the amount that the ions deviated from natural seawater was deduced, which, in turn, allowed the prediction of the ions that could cause toxicity. In looking at Table 7-3, fluoride and calcium are likely causes of toxicity in the Floridan, Biscayne, and recycled Biscayne waters. In addition, bicarbonate, sulfate, and magnesium

Table 7-2 Comparison of raw water and concentrate

Parameter	FL RO Raw mg/L	FL RO mg/L	Fr. MS Raw mg/L	Fr. MS mg/L	MS Recycle mg/L	MS Recycle mg/L	Seawater mg/L	SW RO Conc mg/L
TDS	4953	24462	547	3224	3426	13745	30893	60386
Sr	n/a	n/a	n/a	n/a	n/a	n/a	6.6	12.6
F	1.47	7.13	0.2	0.47	0.5	1.95	0.6	1.33
HCO_3	159	158	288.0	995	995.0	78.2	123.5	78.2
Ca	180	911	95.0	577	577.0	2434	360	686
SO_4	550	3306	46.0	1356	1357.0	8341	2381	4716
Mg	150	759	23.0	228	229.0	966	1136	217
K	n/a	n/a	n/a	n/a	n/a	n/a	349	664
Cl	2461	12161	70.0	202	202.0	849	17002	32292
Na	1401	6910	222.0	56	56.0	236	9463	21642
Ammonia	0.64	3.1	2.0	5.1	5.1	21	1.3	2.9

Ca—calcium, Cl—chlorine, F—fluoride, FL RO—Floridan reverse osmosis water, FL RO Raw—brackish Floridan reverse osmosis raw water, Fr. MS—fresh Biscayne Aquifer membrane system water, Fr. MS Raw—fresh Biscayne Aquifer membrane system raw water, HCO_3—bicarbonate, K—potassium, Mg—magnesium, MS Recycle—recycled Biscayne Aquifer water, Na—sodium, SO_4—sulfate, Sr—strontium, SW RO Conc—seawater reverse osmosis concentrate, TDS—total dissolved solids

Table 7-3 Comparison between salinities of seawater and membrane-filtered concentrates

Parameter	Seawater mg/L	Percent	FL RO mg/L	Deviation fr. Seawater	Fr. MS mg/L	Deviation fr. Seawater	MS Recycle mg/L	Deviation fr. Seawater	SW RO Conc mg/L	Deviation fr. Seawater
TDS	30893		24462		3224		13745		60386	
Sr	6.6	0.021			0				12.6	0.98
F	0.6	0.002	7.13	15.0	0.47	7.5	1.95	7.30	1.33	1.13
HCO_3	123.5	0.400	158	1.6	995	77.2	78.2	1.42	78.2	0.32
Ca	360	1.165	911	3.2	577	15.4	2434	15.20	686	0.97
SO_4	2381	7.707	3306	1.8	1356	5.5	8341	7.87	4716	1.01
Mg	1136	3.677	759	0.8	228	1.9	966	1.91	217	0.10
K	349	1.130							664	0.97
Cl	17002	55.035	12161	0.9	202	0.1	849	0.11	32292	0.97
Na	9463	30.632	6910	0.9	56	0.1	236	0.06	21642	1.17

Note: Values close to 1.0 in "Deviation from Seawater" mean that their values are extremely close to the values present in typical seawater. The light grey highlight displays the likely causes of toxicity, and the dark grey highlight displays the potential for toxicity because of ion imbalance.

might appear to have the potential for toxicity because of ion imbalance, but at lower relative levels. Calcium, fluoride, and potassium in the RO concentrate were consistent with the values present in seawater, while magnesium and bicarbonate were at very low relative levels. More research is needed to determine whether magnesium and bicarbonate deficits would produce significant hazards to the aquatic life.

The following sections will discuss the solutions to problems posed by concentrate ion toxicity and pH imbalance.

Ion toxicity. The methods for assessing this toxicity involve the use of WET tests and have been discussed extensively by Mickley (2000). The report focused on efforts in Florida to deal with the disposal of concentrate from membrane water treatment plants—the majority of which were treating hard groundwater, not brackish water. None were treating wastewater or seawater. Membrane processes are highly effective in removing practically all unwanted particles from water depending on the membrane used. It is these unwanted particles that accumulate in the concentrate that must be discarded. With the information generated by Pillard et al. (2002), methods and information to model the ion imbalance can help identify cost-effective methods of treatment and disposal of membrane concentrate. Interestingly, seawater concentrate has the least potential for toxicity when discharged to ambient ocean water.

The solution to ion imbalance in Hollywood, Fla., was dilution. The concentrate was a minimal share of the total discharge, and WET tests demonstrated that the toxicity was eliminated. However, smaller streams would not present the same opportunities, so chemical adjustments would be necessary to match receiving waters. This could be accomplished, but not without cost and significant ongoing toxicity testing and monitoring. Dilution is a better option where available.

Adjustment of pH. Membrane systems are very particular in terms of water volumes and characteristics of the water, such as pH. Most membranes require incoming water with a pH of approximately 5.5–6.5. This value is acidic. Even after passing through the membranes, the pH of both the concentrate and the permeate is still lower than desired. Modeling with ROSA estimates the concentrate pH; in all cases, the concentrate needs a pH adjustment so that it will not be toxic when it enters the receiving waters. To adjust the pH, in this case, a base must be added. A common chemical for this type of reaction is sodium hydroxide. This basic solution will react with the acidity and cause the pH to rise, thereby creating a more stable, neutral concentrate, which is better for disposal in the environment. The pH must be considered when determining the effects of post-treatment processes on concentrate. The pH affects the concentration and forms certain ions such as the carbonate series and ammonium. Un-ionized ammonia is acutely toxic to marine organisms. High pH also affects marine organisms acutely, which is another problem.

Location Requirements

Surface discharges need to be located near a water body. The water body can be either fresh- or salt water, but it must be moving. RO concentrate from seawater or deep inland sources may have a concentration of approximately 60 parts per trillion (ppt) when it is discarded by the membrane system (as predicted by ROSA). This is almost twice as much as the concentration of typical seawater. Because the majority of aquatic life lives near the bottom of channels and streams, plenty of mixing is necessary so that the emerging concentrate does not just fall and spread out across the bottom of the water body, consequently killing the marine life. To ensure that aquatic life is not harmed, a protocol to dispose of this concentrate must be developed. One suggestion is that the concentrate will only be disposed into the channel during periods when large amounts of mixing take place. The concentrate effluent is released into the channel only at high flow, when there are higher velocities and better dilution and dispersion occur. The major limitation with using high flows is that it only happens periodically (in coastal regions twice daily with tides). Consequently, the storage tanks will need to store the excess concentrate during low tide and low-velocity periods.

Despite the challenges, dilution to attain proper concentration is being pursued. Table 7-4 outlines a series of Florida membrane plants that discharge to surface water bodies. One concept being investigated by the Long Beach Water Department in Southern California is a subsurface dispersion field. The concept is similar in design to an infiltration gallery, in which an engineered sand "filter" is used to disperse the concentrate below the surface. Mixing occurs within the bed, and salinity concentrations are essentially equivalent to background levels by the time the concentrate reaches the seabed floor. Such construction would require low tidal energy to ensure that the dispersion field filter bed remains intact.

The City of Melbourne, Fla., encountered challenges with the disposal of the concentrate at the original discharge location into the Eau Gallie River. To overcome these challenges, the city commissioned studies including a concentrate toxicity evaluation to assess the ion imbalance; a three-dimensional numerical modeling of the river; a diffuser evaluation to determine the mixing zone size to meet the surface water standards (for gross alpha and radium); an evaluation of the concentrate post-treatment; and an assessment of the concentrate transfer pumps. The monitoring requirements included manatee observation and temperature monitoring at the discharge location to ensure that the concentrate did not affect the manatee behavior in winter.

Tampa Bay Water disposes of concentrate into Tampa Bay by mixing it with large volumes of cooling water from the neighboring power plant to limit toxicity of the concentrate. In Perth, Western Australia, wastewater is discharged to the ocean with vertical dispersion systems to limit toxicity.

Construction Methods

Surface water construction is based on extending a pipe to the receiving water (see Figure 7-3). The cost is location-specific. The pipe can be installed by cut and fill, trenching, or directional drill. All are standard construction methods that are well understood by those in the field.

However, for ocean outfalls, laying the pipe on the surface of the seafloor is not enough. Trenching is impossible underwater in deep zones, so armoring (with articulated mats or riprap) is required. Tunneling may be preferred (similar to directional drilling). However, tunneling or armored pipes can cost more than $10,000/lin ft.

Table 7-4 State of Florida potable water reverse osmosis and nanofiltration systems that discharge to surface waters or ocean outfalls

Plant	Startup Year	Capacity (mgd)	Type of Process	Water Recovery (%)	Waste Stream Dilution Factor	Concentrate Disposal
City of Cape Coral	1976	17	RO 2-stage	80	4.3	Surface saltwater lake
City of Vero Beach	1991	2	RO 2-stage	85	0.4	Surface water – Indian River Lagoon
City of Melbourne	1995	6.5	RO 2-stage	85	1.1	Surface water discharge
City of Palm Coast	1992–2001	6.4	NF 2-stage	85	1.1	Surface water discharge – freshwater canal

Note: Information was collected directly from water treatment facilities staff.

Figure 7-3 Surface water discharge

Operations and Maintenance Requirements

Maintenance of freshwater piping will be similar to that of a distribution pipe or intakes for surface waters. The potential for damage is limited as the pipe is generally not under pressure. Tunnel and ocean outfalls are far more difficult to repair and require special equipment and crews.

Deep-Well Injection

Deep-well injection, a viable method of disposal, is covered in detail in Chapter 4.

Brine Concentrators

Brine concentrators are covered in Chapters 4, 5, and 9. The most significant issue with brine concentrators from a regulatory standpoint is the management of the solids. Ensuring that the solids will meet a nonhazardous characteristic is critical in selecting the appropriate brine concentrator. Overall, if the system generates a solid or concentrated waste, it must be tested to determine the proper classification of the material and how it must be handled by the municipality.

Inland Desalination and Public Outreach

Inland desalination is proving to be a viable way to develop reliable alternative water resources for many communities. Yet, like all water projects from reuse to dams, inland desalination can be met with a surprising degree of public opposition. In their comprehensive study of desalination plants in the United States and around the world, Millan et al. (2015) found that nearly all of the projects surveyed encountered "some opposition, and often strong opposition."

The primary issues of contention ranged from the remarkably well informed—such as the environmental impact of brine discharge and disposal of concentrates, energy usage/carbon footprint/cost, and sustainability (short RO membrane shelf life)—to the highly practical, such as construction and operational costs, visual impact and plant aesthetics, and noise emissions during plant operation. Not surprisingly, the environmental impact on marine and other wildlife (entrapment, impingement, and entrainment) was also a major concern (Millan et al. 2015); these authors cited sophisticated opposition campaigns directed by several well-established environmental organizations, particularly in places such as California where the public awareness and concern over environmental protection and preservation is particularly high.

Finally, other opposition concerns included perception of population growth inducement; construction inconveniences; not doing enough to conserve existing water resources; and failure to explore other options (Millan et al. 2015). The authors warn that failure to proactively meet such wide-ranging and often well informed and organized opposition to a desalination project can have enormous consequences that have included substantial concessions or critical project redesign and have even resulted in the cancellation of some much-needed projects.

All of this makes a well-thought-out and effectively orchestrated program of public outreach a critically essential communication tool to achieve a successful outcome (Millan et al. 2015).

Desalination may indeed be the perfect antidote for the seemingly inevitable need for new and sustainable water sources many communities are facing today or will face in the future. When it comes to the public and water-source changes, however, the challenge often lies in communicating the risks—and the benefits—involved before the permitting process for desalination projects. It is not that ingenuity to increase water portfolios, such as the implementation and use of the relatively new technology of inland desalination, must be met with skepticism; it's more a factor of bringing the public along from the beginning to ultimately gain their support.

Generally, when it comes to emerging technologies, such as desalination, it is best to communicate any risks or community impact early on, rather than try to provide a solution to the public in the future. Often, it means communicating the risks to the public proactively in an effort to avoid public outrage because of rumors and misinformation.

Millan et al. (2015) grouped these risks into three general categories:

- the most significant challenges to public acceptance;
- bureaucratic regulatory challenges from federal, state, regional, and local government interests; and
- broader, overarching challenges that stem largely from the public's understandable lack of knowledge (and often that of their political leaders) about the crucial economics and the basic fundamentals of their critical water systems.

Consumers do not spend their days thinking about the water that comes from their taps, and they are not aware of what it takes to make that happen daily. It is against this backdrop that implementation of new and emerging technologies such as inland desalination must be attended—and preceded—by appropriate efforts to educate the public about the critical aspects of their water resources and the true cost of clean water. Educational topics include the cost of desalination, energy demands, project location, and construction impact, but they should also address environmental and wildlife impact. It is also critical to educate the public regarding projected future population and infrastructure growth and the need to ensure long-term sustainability.

The political and governmental challenges to implementation derive directly from endeavoring to gain regulatory approvals from myriad agencies or departments. The regulatory gauntlet to final approval is often a burdensome and expensive process that can take years to complete. In some cases, it is also fraught with potential politically motivated peril or infighting. Yet, understanding the political pitfalls makes clear that an effort to educate and work with key regulators and government officials will go a long way toward creating bipartisan support that can help move desalination projects through an arduous approval process.

For example, Millan et al. (2015) reported that cooperation and open, educational communication with government regulators resulted in turning some officials into "project champions or advocates that were instrumental in gaining support for (proposed) projects." And while profound and consequential strains on existing water resources and infrastructure system facilities—such as supply issues because of excessive drought or damage from severe flooding—have often played the most significant role in winning widespread public and political support for desalination projects, these are random factors that cannot be relied upon.

Perhaps the greatest overall challenge is getting people to understand the project-specific circumstances that become the drivers for implementing new water supply options such as inland desalination. Such circumstances may be climatological or simply a result of community and population growth. As noted previously, consumers tend to take reliable water supply for granted. Therefore, the success of any new water initiative depends on educating the public, elected officials, and the media about the benefits of developing new water sources, as well as demonstrating how new technologies, such as desalination, fit into a community's wider water supply portfolio.

As previously noted, the public lacks knowledge about the economics and basics of water systems. Therefore, providing a basic information "primer" about water systems and supply challenges—before introducing new and potentially controversial options—is a key component in the process. Such a primer should address the typical issues of concern among opponents cited earlier, such as cost, energy demands, marine-life impact, plant location, and aesthetics.

Recognizing the Problem

As the desalination process becomes more financially feasible, it is important to reveal the initial problem of water source sustainability—not just the solution. It makes it incumbent upon the water industry to educate the public on the benefits of using inland desalination and the basics about finding sustainable water supplies (WRD 2011). It falls upon the water provider to frame the debate, first by defining the problem and then by presenting the potential solutions. For the experts, the challenge can be knowing when to begin the dialogue with the public to discuss the problem and potential solutions.

What the utility might define as a problem might be viewed differently by the public. This can be a time-consuming challenge, even for the communication experts. In most cases, the public is not aware of the ongoing need to ensure a continuous, sustainable water supply and will resist a utility's new proposals. In addition, few people pay attention to data they deem nonurgent; fewer still take the time to educate themselves and consider alternatives. In a 2008 survey of water customers conducted by the West Basin Municipal Water District, which serves about one million people in 17 cities and unincorporated areas in Los Angeles County, water ranked last on the list of important issues facing consumers. A year later, concern over water had risen to second in importance to the district's customers. The survey also revealed that ocean-water desalination had become one of the top three ways people envisioned increasing water portfolios (WRD 2011).

Even the most elegant solution is doomed to fail without public assessment and support, especially when it involves an emotionally charged issue such as water supply (WRD 2011). Most communities are reticent to assume financial and health risks on the testimony of a "faceless" water provider. Hence, it is essential to allay potential outrage by involving people and educating them at the outset of the project and permitting process (Millan et al. 2015).

Toward Solutions

Once the public acknowledges that a problem exists, such as an impending water shortage, gaining the public's support with potential solutions often becomes easier.

One caution is the public will often attempt to educate themselves on the issue by browsing the Internet. This is where the project's communications team needs to ensure that urgency is rewarded with good information on the pertinent project website and also to understand that the utility's website may not be the only source of information for the public. Members of the public have been known to surf a variety of websites, looking for those they perceive to be objective and highly informed. For this reason, it is advisable to provide links on the utility's website to alternative sources of credible information for its customers and to use third-party validation when possible when presenting statistics and pertinent information.

The difficult task of responsible decision-making is less burdensome when citizens are well informed. Providing clear, accurate, and unbiased information, regardless of the source, steers the public away from unreliable sources of information. It also steers them away from actively oppositional data that might result from overly intensive, poorly defined searches.

In the end, however, some blend of fact and perception—including misperception—may be used by the public in their decision-making process. Often people will consider more than one solution, so the utility should not oversell anything as the "perfect" solution; at the same time, the utility should be prepared to reveal why other solutions are not deemed to be viable (Millan et al. 2015).

Public Risk

No water treatment solution is 100 percent safe or without some risk, and inland desalination is no exception. Every solution entails some element of compromise and a list of tradeoffs. So, when the utility steps into the public relations arena, it is helpful to empower people by validating their contribution, no matter how small. It also means revealing the risks as the public's imagination is often far worse than the reality.

Common concerns of the public regarding the use of inland desalination include likelihood of rate increases, impact on the community, health impact, environmental issues, and energy requirements. What research continues to conclude is that with any inland desalination project, no matter how unique, proactive public outreach becomes essential to the overall success of the project (Millan et al. 2015).

Communication Before and After Implementation of the Solution

Support for a solution to a complex and controversial water-supply problem is a choice that can easily shift several times over the course of weeks or even years. A steady stream of information throughout the process serves at least two purposes. It affirms and solidifies public opinion and choices, but more importantly, it supplies information to "opinion-leaders"—people who have given their support and now may be expected to sway other potential supporters to the utility's side.

The important thing to gain from this strategy is that the public's understanding of the problem and solution is not the end of the journey—it is the beginning. Keep the information flowing throughout the process, even after they are on board with the utility's "best" solution and especially after they have reached into their pocketbooks to support it. Also, people move into and out of neighborhoods. Just because the utility has informed people at one point does not mean "everyone" has heard it at later points in time. Document public approval with supportive quotes on the utility's website, or have testimonials woven into other communication tools. These are expressive "snapshots" of communication success. Public opinion—the viewpoints of friends, neighbors, and colleagues—will serve the utility well at the beginning, middle, and end of the public engagement journey. This is where focus groups, surveys, meetings, and research can prove invaluable in the communication toolbox (Millan et al. 2015).

A supportive public—educated, engaged, and free of misinformation—is the ideal endpoint of the outreach journey. Constructive, two-way interaction is a tool which can be wielded effectively to establish relationships, build trust, and shape the decision-making process by positioning the water industry experts and leaders as people who honestly care about the common good.

Basic Rules of Engagement

Develop a communication plan. From the beginning, develop a strategic communication plan detailing timeframes, budgets, and how to measure its success. Communicate early and often to the community, and be prepared to engage critics and address concerns and fears with solid, documented information. Effective plans should be flexible and active; think of them as a verb, rather than being a noun that sits on the sidelines. Because things can go wrong, include a section in the plan on dealing proactively with a potential crisis. And be prepared to proactively report any mishaps—which are bound to occur—along with a solution for avoiding similar problems.

Take a snapshot. Every public water project is unique and thus there is no perfect process for public outreach. Yet, understanding the people that ultimately can support the project can help to ensure the overall success of the communication plan. That is why one of the first things is to engage with the public, stakeholders, and elected officials to learn viewpoints, cultures, and what people understand about water supplies and desalination. This is also a time to exchange information and validate people's concerns. Transparency is key because it builds trust. The notion of a "snapshot" is no simple metaphor here, because a snapshot shows things as they are at a given point in time. Some methodologies to consider using are focus groups, surveys, citizen committees, and one-on-one dialogues.

Craft the message. Any messaging to the public should be targeted to specific audiences; it should recognize alternative viewpoints, yet be thoroughly consistent. It tends to be most effective when the messaging begins upstream, at the permitting process. It is important to reveal all the risks, environmental and economic, and not just the benefits of inland desalination when crafting the message. Have a team in place ready to answer questions and concerns as quickly as possible. Always make communication and community outreach paramount in the comprehensive game plan.

Develop communication materials. A useful ingredient of public outreach can be hard-copy materials, which can be as simple as a one-page handout. The information should be easy to understand and inviting to read, with easy references or links to where people can get more information, such as a project-specific website (Millan et al. 2015). It may be useful to provide literature profiling successful inland desalination projects across the United States and around the world.

Inform the public. When people know the risks as well as the benefits of the desalination process, they are likely to be less alarmed. When the National Institute for Chemical Studies surveyed people living near a chemical plant, the results revealed residents that had watched a video about how to take shelter in the event of an emergency perceived a lower risk of living near the chemical plant. Those that did not watch the video had a much higher perception of the risks of living near the plant. While there are also legal issues that need to be addressed with a public outreach program, it is better to have more information on the risks and the mitigation steps for a particular project than to provide only the benefits. People that are informed of the risks, and not just the benefits, are more likely to provide support for the implementation of a desalination project. Importantly, get the facts straight; line up the best experts available to authoritatively and definitively answer opposition questions. Be available to attend regular community events; doing so marks the communications team as fellow members of that community, with the same concerns and needs.

Embrace the media. Be proactive with the media. Like the public, the local media should be brought into the mix early on. Include them on meeting notices, send them press releases, provide them with tours, and keep them updated with the facts. The better the relationship that is established with the media (like with the public), the higher the likelihood of at least getting to tell the desalination project's side of a story should negative press occur. Try to gain support from key groups, or perhaps from community civic or business leaders—even celebrities—whom the people know and tend to trust, to popularize the desalination project. There might be a popular champion who is highly relatable—and thus potentially very influential—to the community.

Provide informative tours and events. If the municipality provides tours and explanation sessions on the permit and what they are trying to achieve, it likely will provide a sense of security that there is nothing to hide and that the municipality does have the public's interest in mind during this process. Be sure to keep local elected officials, business leaders, and community leaders informed about meetings and consider inviting them personally on tours. Be creative and innovative about the approach to community meetings. In their study, Millan et al. (2015) found that while traditional public meeting forums were helpful, more productive and substantive dialogue often resulted from "an open house, information-station style meeting where participants can talk one-on-one with project team members."

Provide lab testing and results. It may be important to provide comparisons of scientific lab data. Providing data in spreadsheets or in graphs, while informative, will not be useful to the public. That information can be provided as a backup to the overall information sheet, but it is better to provide a comparison that relates to some feature that the public generally has some knowledge and understanding of, in a manner they can trust. Overall success of desalination projects often depends on providing the public with useful information early in the permitting process (WRD 2011). Make sure the communications team is prepared to answer noncentral concerns such as environmental and wildlife impact, greenhouse gases, or noise pollution. Having ready responses to these concerns will demonstrate that the team has been meticulous in broadly assessing project impact on all possible levels.

Use interactive tools. There are numerous interactive tools that can be used to provide information to the public. An example would be the use of a three-dimensional video of an injection well. This can demonstrate how the injection well can be constructed to protect groundwater because of specialized, site-specific well construction techniques that take into account the area's geology. The utility should communicate that the hazards of a particular operation have been identified and fully considered. In addition, if concerns are voiced, be ready to state honestly that, after receiving comments, the project team plans to review the work and that it fully expects the system will be made more secure thanks to the public's probing questions. This allows for direct impact on the process through the public's active participation. When people feel like they are a part of the process, they naturally tend to be increasingly supportive of the project's plans and goals.

Use social media and websites. These communication tools can be useful in reaching a certain segment of the population and help to modify public perception of a project. However, be aware of the issues associated with information that is not correct or is misleading. Consider developing a project-specific website and use major social media platforms such as Facebook, YouTube, and Twitter (Millan et al. 2015). Should rumors or misinformation be posted on the Internet, do not try to rebut or refute or be defensive;

rather, it is better and more effective to continue to get the correct information out, in abundance, as the best means of educating the public.

Keep employees updated. Internal staff should be kept updated and aware of the facts as the project unfolds. Develop a list of "Frequently Asked Questions" (FAQs) that can be used to inform both internal staff and the public. Hold internal customer service meetings to ensure the messages going out on emails and the phones are consistent with the overall messaging of the project (Millan et al. 2015). Make sure communications officers or teams are not buried under bureaucratic layers of the organization; they must be the most accessible people on the front lines of information dissemination. If the city or town has a communications officer charged with keeping the public informed, keep him or her fully informed of ongoing developments. Working closely with city communications officers is often a subtly effective way of making "city hall" a collaborative member of the project team and a proponent of the project.

In Summary

It is almost inevitable that any permitting activity will bring up concerns with the public. Accordingly, it is imperative to communicate early in the permitting process—both the benefits and, perhaps even more crucially, the risks and ramifications that are associated with inland desalination—and to keep the public integrated. Concerns about the need to ensure water resources for the long term and to build and maintain new and innovative state-of-the-art delivery systems are not top of mind for the general public; consumers take reliable water for granted. Yet, experience has shown that not only do they wish to be informed, there are often important segments of the community that want to have the opportunity to provide active and thoughtful input into the project, from conception and design through development, construction, and completion.

While every inland desalination project is unique, it is important to gauge public viewpoints and to start the outreach process in the early stages of the permitting. Doing so will appreciably increase the chances for project success.

REFERENCES

Ahmed, M.; Shayya, W.H.; Hoey, D. and Al-Handaly, J. 2001. Brine Disposal From Reverse Osmosis Desalination Plants in Oman and United Arab Emirates. *Desalination,* 133(2):135–147.

AWWA. 2008 (3rd ed.). M21, *Groundwater*. Denver: AWWA.

California Department of Water Resources. 2013. *California's Groundwater Update 2013*. https://water.ca.gov/-/media/DWR-Website/Web-Pages/Programs/California-Water-Plan/Docs/Update2013/GroundwaterUpdate/Californias-Groundwater-Update-2013--A-Compilation-of-Enhanced-Content-for-California-Water-Plan-Upd.pdf (accessed Feb. 19, 2019).

Dawoud, M.A. and Al Mulla, M.M. 2012. Environmental Impacts of Seawater Desalination: Arabian Gulf Case Study. *International Journal of Environment and Sustainability,* 1(3):22–37.

Del Bene, J.V.; Jirka, G.; and Largier, J. 1994. Ocean Brine Disposal. *Desalination,* 97(1–3):365–372.

DiNatale Water Consultants. 2012. *Water Partnerships: An Evaluation of Alternative Agricultural Water Transfer Methods in the South Platte Basin*. http://dinatalewater.com/pdf/water_partnerships.pdf (accessed Feb. 19, 2019).

Finley, B. 2015. A Pilot Project Aims to Extract More H_2O From Already Treated Wastewater. *Denver Post,* July 16.

Florida Department of Environmental Protection (FDEP). 2004. *Technical Guidance Document for Conducting the Florida Department of Environmental Protection's Protocols for Determining Major-Seawater-Ion Imbalance Toxicity (MSIIT) in Membrane-Technology Water-Treatment Concentrate.* http://publicfiles.dep.state.fl.us/dear/labs/biology/miscpubs/msiit_final.pdf (accessed Dec. 31, 2018).

Kaunisto, D. and Mullenix, R. 2006. *A Refreshing Water Source.* http://americancityandcounty.com/water/government_refreshing_water_source (accessed May 3, 2018).

Mickley, M. 2003. *Review of Concentrate Management Options.* texaswater.tamu.edu/readings/desal/concentratedisposal.pdf (accessed May 3, 2018).

Mickley, M. 2009. Desalination and Water Purification Research and Development Program Report No. 155: Treatment of Concentrate. Washington: US Bureau of Reclamation.

Mickley, M. 2000. *Major Ion Toxicity in Membrane Concentrate.* Denver: AWWA Research Foundation and AWWA.

Mickley, M.C.; Hamilton, R.; Gallegos, L.; and Truesdall, J. 1993. Membrane Concentrate Disposal. Denver: AwwaRF and AWWA.

Millan, M.; Tennyson, P.A.; and Snyder, S.A. 2015. Model Communication Plans for Increasing Awareness and Fostering Acceptance of Direct Potable Reuse. Alexandria, Va.: WateReuse Research Foundation.

Pillard, D.; DuFresne, D.L.; and Mickley, M.C. 2002. Development and Validation of Models Predicting the Toxicity of Major Seawater Ions to the Mysid Shrimp, Americamysis Bahia. *Environmental Toxicology and Chemistry*, 21(10):2131–2137.

South Florida Water Management District. 2009. *Water Desalination Concentrate Management and Piloting.* https://www.sfwmd.gov/sites/default/files/documents/waterdesalinationconcentrate mgmtpiloting-dec09-carollo.pdf (accessed Nov. 7, 2018).

Squire, D. 2000. Reverse Osmosis Concentrate Disposal in the UK. *Desalination*, 132(1–3):47–54.

TCEQ (Texas Commission on Environmental Quality). 2007. Chapter 218: Brine Evaporation Pits. Austin, Tex.: TCEQ.

Van Voorhees, R.F. 2007. Underground Injection Permitting. Proc. Nuclear Regulatory Commission and the National Mining Association Uranium Recovery Workshop, May 2007, Denver.

WRD (Water Replenishment District of Southern California). 2011. *Engineering Survey and Report.* https://www.wrd.org/sites/pr/files/WRD_ESR2011_May4_Final_Report_web.pdf (accessed Feb. 21, 2019).

Chapter 8

Case Studies

Howard E. Steiman, Qigang Chang, Charlie (Qun) He, Phil Locke, Ryan Popko, Jim Hogan, Fred Greiner, Jason Yoshimura, Bill O'Neil, Greg D. Wetterau, Shane Walker, Malynda Cappelle, Tom Seacord, Graham Juby, and Brandon Yallaly

This manual contains eight case studies to illustrate several of the approaches discussed herein. Table 8-1 identifies the facilities included as case studies and the technologies they demonstrate.

CITY OF CHANDLER OCOTILLO BRINE REDUCTION FACILITY

The City of Chandler, Ariz., operates an industrial process water treatment facility, originally named the City of Chandler Reverse Osmosis Facility (CHRO). The facility treats brine generated by Intel Corporation's internal RO water treatment systems at its Ocotillo Campus and recovers water for reuse and recharge. A concentrate management system upgrade project was completed in spring 2014, after which the CHRO was renamed the Ocotillo Brine Reduction Facility (OBRF).

As Figure 8-1 shows, before the modifications in the spring of 2014, the CHRO consisted of the following major components:

- Low-pressure feed pumps: Waste brine from the Intel Ocotillo campus flowed by gravity into two wet wells. Each wet well was equipped with three vertical turbine pumps to convey the waste brine from the wet wells to the downstream treatment processes.
- RO pretreatment systems: Water from the low-pressure feed pumps was chemically treated with acid and scale inhibitor to prevent sparingly soluble salts such as calcium carbonate from precipitating in downstream RO processes. Following chemical treatment, the existing feedwater was pumped through 10-µm cartridge filters to remove solids before the RO treatment systems. A microfiltration

Table 8-1 Case study facilities

Facility	Owner	Concentrate Management Approach
Ocotillo Brine Reduction Facility	Intel Corporation, Chandler, Ariz.	Zero liquid discharge utilizing softening for enhanced recovery with a brine concentrator and evaporation ponds
Water Treatment Plant No. 2	City of Palm Coast, Fla.	Zero liquid discharge using lime and soda ash softening and recycling
Fort Irwin, Calif.	US Army	Deep-well injection
City of Fargo Wastewater Effluent Reuse	City of Fargo, N.D.	Surface water discharge
Kay Bailey Hutchison Brackish Groundwater Desalination Plant	El Paso Water, El Paso, Tex.	Current: deep-well injection; future: enhanced concentrate recovery through recovery of saleable minerals employing treatment processes such as softening, nanofiltration, and electrodialysis
Confidential, Calif.	Confidential, Calif.	Zero liquid discharge using brine concentrator and evaporation ponds
Laguna County Sanitation District, Santa Maria, Calif.	Laguna County Sanitation District	Concentrate volume minimization and deep-well injection
Chino Desalter Authority Concentrate Reduction Facility	Chino Basin Desalter Authority, Calif.	Enhanced concentrate recovery and brine line disposal

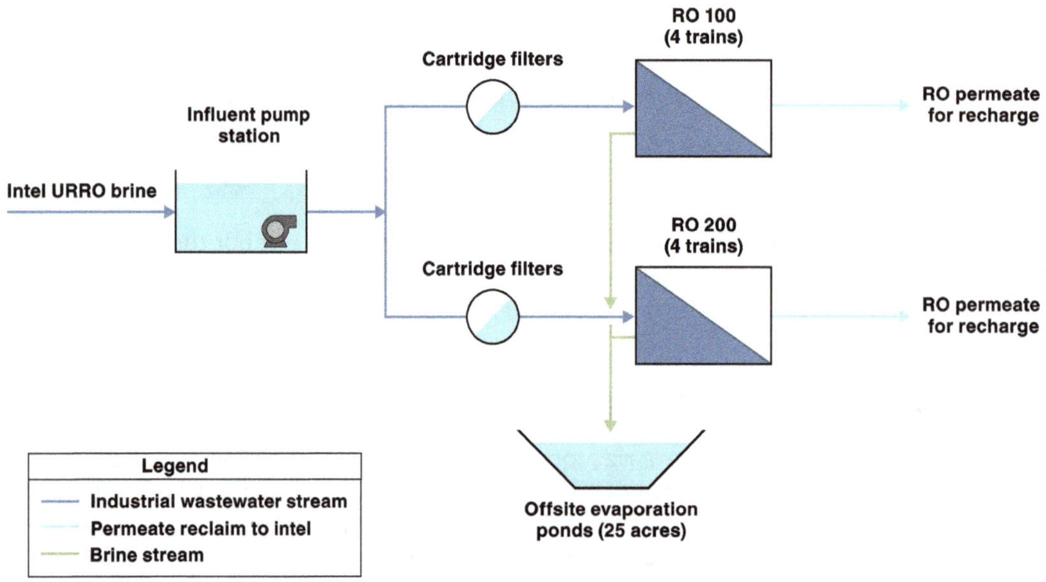

Figure 8-1 CHRO process flow schematic (prior to spring 2014)

(MF) system was installed upstream of one of the two RO treatment systems to provide a more robust pretreatment of wastewater from Intel. The MF system was decommissioned shortly after the CHRO facility started up because of a change in the quality of the wastewater received from Intel.

- RO systems: CHRO had two independent RO treatment systems. Each consisted of four RO treatment trains. Each train included a dedicated high-pressure pump, pressure vessels that house the membrane elements, and instrumentation and control equipment. The RO systems were designed for approximately 70 percent recovery.

- Product water pump station: Treated water from the RO systems was collected in a wet well and transferred to offsite aquifer recharge wells.
- Chemical feed systems: Sulfuric acid, scale inhibitor, and sodium hypochlorite were used in various phases of the treatment process. The sulfuric acid system consisted of a steel storage tank and metering pumps that fed acid to the raw water downstream of the wet wells used for receiving Intel's wastewater. The scale inhibitor system consisted of storage totes and metering pumps that fed inhibitor into the raw water downstream of the acid addition point. Sodium hypochlorite was stored in a double contained tank and dosed via metering pumps into the treated RO permeate before being sent to the aquifer recharge wells.
- Clean-in-place (CIP) system: The CIP system consisted of two solution tanks, a CIP recirculation pump, and a cartridge filter. The system was used to remove foulants and scale from the membranes without removing them from the pressure vessels.
- Brine evaporation ponds: Brine from the RO facility was conveyed under residual pressure through a dedicated 8-in. brine line to an offsite evaporation pond facility located 6–7 mi east of the CHRO. The facility consisted of four lined ponds with a total surface area of 25 acres. The CHRO was also equipped with four acres of onsite brine ponds. However, these ponds were not part of the CHRO operation and were primarily designed to receive blowdown from a brine concentrator located on the Intel campus (treating another industry flow). From a meteorological perspective, CHRO is located in a net evaporation region, making evaporation a feasible brine disposal alternative.

Process Challenges

Because of planned expansion at Intel, the waste brine volume was expected to increase and to subsequently exceed the capacities of CHRO and the offsite evaporation facility. To address the anticipated brine volume change, Intel awarded a project in July 2012 and designated it as the OBRF. The purpose of the OBRF was to design and construct major process modifications to accommodate the additional brine volume, maximize the system recovery, and reduce the overall brine flow to the evaporation ponds. The OBRF also provided another benefit: It would reduce the consumption of the City of Chandler's potable water at Intel's Ocotillo campus by supplying high-quality permeate that could be reused for cooling water and other onsite purposes.

Maximizing membrane system recovery and thereby minimizing concentrate volume was a project goal. However, maximizing membrane recovery could not be achieved without first treating the concentrate to remove scale-forming minerals that inhibit membrane performance (i.e., calcium, magnesium, silica, and TOC). To facilitate higher RO system recoveries, chemical precipitation using lime softening followed by granular media filtration and weak acid cation exchange were selected as RO process pretreatment steps. The chemicals selected to raise pH were $Ca(OH)_2$ (lime) and NaOH (caustic soda). CO_2 (carbon dioxide) and caustic soda were selected in lieu of Na_2CO_3 (soda ash) to provide carbonate alkalinity.

The proposed process strategy was to use softening to effectively remove calcium to low levels (less than 30 mg/L as $CaCO_3$). Magnesium chloride can be added to enhance silica removal, although the raw water currently has sufficient magnesium to remove silica to less than 20 mg/L as SiO_2. After hardness polishing to less than 0.1 mg/L total hardness with ion exchange, the downstream RO processes are capable of achieving 90 percent recovery, versus the original CHRO recovery of 70 percent.

High TOC concentrations could result in organic fouling of the RO membranes. Softening processes, aided by ferric chloride and/or polymer, remove more than 50 percent of the incoming TOC. Maintaining a high pH (around 10) through the RO system helps to inhibit the biological fouling. Moreover, because the solubility of organic compounds would be much higher at high pH, such an approach would serve to reduce organic fouling potential by keeping the organics in dissolved form.

A brine concentrator was included in the system to further reduce the RO brine volume by approximately 70 percent. The Intel Ocotillo Campus was equipped with an existing brine concentrator unit, which had been out of service for several years. The owner decided to relocate this brine concentrator from the Intel Campus to the OBRF facility and repurpose it for the proposed process. Because of the long lead-time of a new brine concentrator, it was determined that only redundant mechanical components (i.e., blowers, pumps, boilers) and controls would be provided (in lieu of a complete redundant brine concentrator system).

OBRF Process Overview

Since its completion in the spring of 2014, the OBRF consists of the following major components:

- Softening treatment facility: Waste brine is pumped to two high-rate clarifiers and stabilized using a conventional lime softening process. Caustic soda, lime, and carbon dioxide (when required to supplement alkalinity) are added to raise the pH to above 11.3 and precipitate calcium, silica, barium, and other sparingly soluble ions that limit the system recovery. Ferric chloride, polymer, and magnesium chloride can also be added as needed to accommodate fluctuations in the influent water quality. Softened concentrate from the high-rate clarifiers is filtered through four granular media filters. The nominal design capacity of this facility is 2.8 mgd.
- High-Efficiency RO (HERO) system: Water from the granular media filters is discharged to a high-efficiency RO facility that includes a weak acid cation exchange polishing process and an innovative four-stage, high-recovery RO with energy recovery. The design recovery is approximately 90 percent.
- Chemical storage and feed facility: Various chemical storage and feed systems are provided at the OBRF, including lime silos and associated batching systems, caustic soda, carbon dioxide, magnesium chloride, ferric chloride, polymer, hydrochloric acid, sulfuric acid, and scale inhibitor.
- Solids handling facility: Lime sludge removed from the high-rate clarifier is collected in a thickened sludge storage tank. Thickened sludge is further dewatered in the solids dewatering facility using three belt filter presses. Dry lime sludge is disposed of at a landfill. Filter backwash waste is collected in the backwash waste equalization tank and pumped back to the headworks of the high-rate clarifiers.
- Brine concentrator: A brine concentrator is provided to further reduce the RO brine volume by 70 percent. As mentioned previously, this unit was refurbished and upgraded by the manufacturer and repurposed to meet the design requirements of the OBRF project before being relocated to the site.
- Brine evaporation ponds: The four-acre onsite brine ponds were also repurposed to receive brine concentrator blowdown as a flow-through storage pond, where the blowdown cools down and further reduces the volume through evaporation. The onsite pond is equipped with a new brine transfer pump station, which transfers the brine concentrator blowdown to the existing 25-acre offsite pond located east of the OBRF facility.

Figure 8-2 depicts a general process flow diagram for the OBRF after spring 2014.

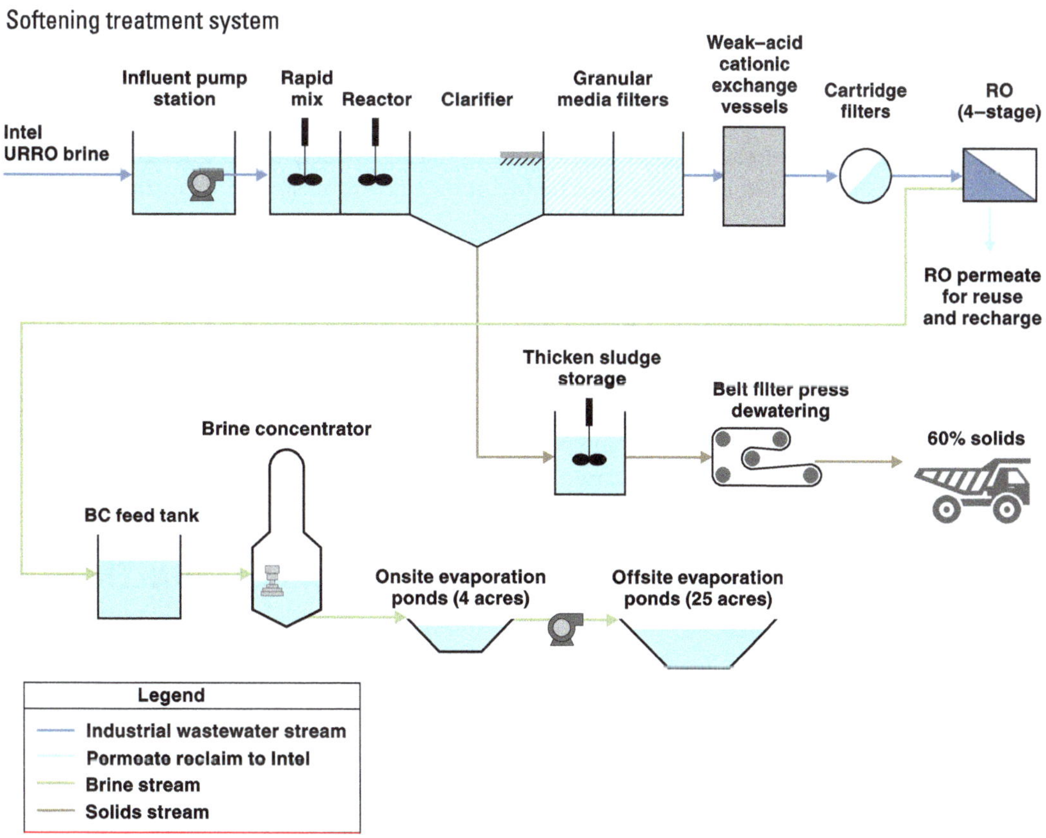

Figure 8-2 OBRF process flow schematic (after spring 2014)

Process Permitting Requirements

To accommodate repurposing the onsite and offsite evaporation ponds, the facility's existing Aquifer Protection Permit (APP) was amended through the Arizona Department of Environmental Quality (ADEQ). The approval process took approximately 250 days to complete.

To support operation of the brine concentrator, an Authority to Operate under the General Permit to Operate and/or Construct for Fuel Burning Operations was obtained from the Maricopa County Air Quality Department. This permit allows the OBRF to operate the natural gas-fired boilers that are operated occasionally with the brine concentrators.

For the safe storage and handling of hazardous chemicals and quick response in an emergency, the OBRF is enrolled in the Hazardous Material Program administered by the City of Chandler Fire Department (which was updated as part of the OBRF project). A nonhazardous liquid waste vehicle permit was obtained by the contract hauler from Maricopa County Environmental Services Department for hauling dewatered lime solids generated by the treatment process.

Project Delivery

The OBRF project was delivered under a design-build procurement approach (DB). This offered several benefits for expediting the project, including improved communication between designers and constructors, early involvement of the constructors in constructability review, and proactive equipment and material procurement and scheduling. Such an arrangement can also reduce the likelihood of disputes between the designer and the constructor from the owner's perspective. With a DB, the interface with the owner between the designer and the constructor is seamless as they are both part of the same project-delivery entity.

The design process started in July 2012. The pre-procurement package and proactive permitting coordination efforts allowed the constructors to start excavation and early procurement of critical equipment and material as early as August 2012. The design was completed within five months. The facility was commissioned in spring 2014.

Operation Challenge

The key operation challenge facing the OBRF is variable feedwater quality. The design of the OBRF includes several features that allow the process to respond to this challenge. The system includes multiple chemical feed systems to manage changes in alkalinity, hardness, silica levels, and sludge quality. This flexibility has enabled the OBRF to respond to seasonal water quality changes and has allowed for treatment of an additional waste stream (cooling tower blowdown) that it was not originally designed to handle.

CITY OF PALM COAST ZERO LIQUID DISCHARGE REVERSE OSMOSIS CONCENTRATE MANAGEMENT

Facility Summary

The City of Palm Coast, located approximately 60 mi south of Jacksonville, Fla., owns and operates Water Treatment Plant Number 2 (WTP No. 2), a facility that uses low-pressure RO (LPRO) for treatment of its groundwater source supply. WTP No. 2 originally had a permitted capacity of 6.38 mgd that consisted of 1.58 mgd of raw water bypass and 4.8 mgd of LPRO permeate. The RO membrane system operates with a recovery of approximately 80 percent. This results in a 1.2 mgd concentrate flow. The LPRO concentrate stream was originally discharged to a surface water body using an extended mixing zone as a licensing basis to achieve applicable water quality compliance standards. Because of subsequent regulatory changes affecting the mixing zone, the City implemented a unique ZLD approach for concentrate management.

The ZLD process treats the 1.2 mgd LPRO concentrate flow and conditions it so that it can be used as part of the city's drinking water supply after lime/soda ash softening, pH adjustment, ultrafiltration (UF), and disinfection steps. With this approach, the ZLD process allows the city to recover 98.5 percent of the LPRO concentrate stream. The ZLD process treats the concentrate such that, when blended with the LPRO permeate and bypass water, the blended, composite stream meets all primary and secondary drinking water standards, achieves city goals, and increases the capacity of the facility by 1.2 mgd without the need for additional supply groundwater sources. The upstream LPRO serves to enhance the efficiency of the ZLD's softening process by concentrating the hardness in the raw well water so more of the hardness is removed by the softening process.

Membrane System Overview

The city's WTP No. 2 RO membrane system consists of five RO trains and a CIP system. Each two-stage RO train has seven 8-in. Hydranautics ESPA1 (energy saving polyamide) elements per pressure vessel. Each train has 201,600 ft^2 of membrane area with an average membrane flux of 14.9 gfd. The system typically operates at 83–84 percent recovery with a first-stage feed pressure of 170–185 psi. Flux balancing between the first and second stage is controlled by throttling the first-stage permeate with a backpressure of 30–40 psi.

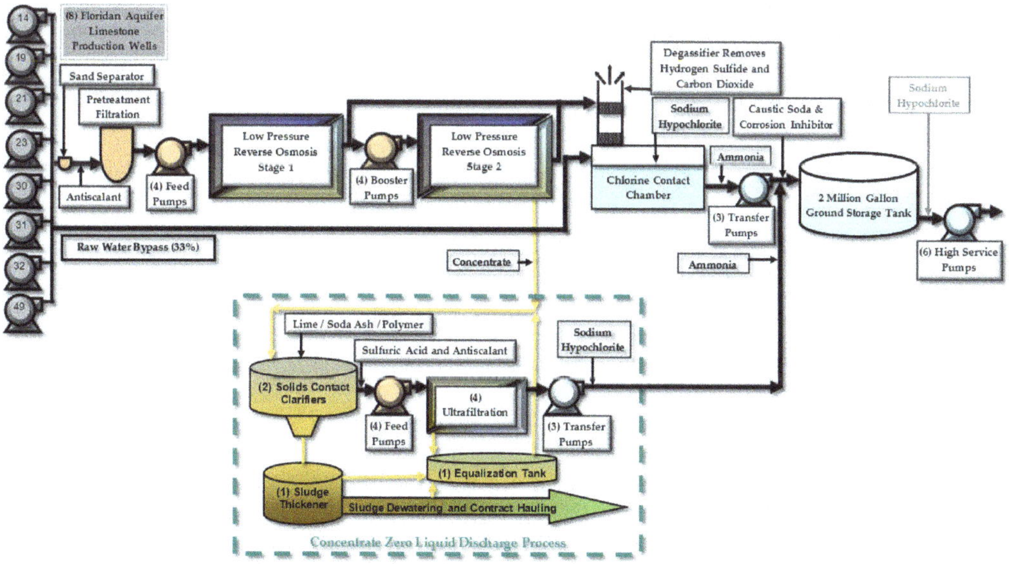

Figure 8-3 ZLD process schematic

RO Concentrate Management and Permit

To select the ZLD process, the city evaluated 12 concentrate management alternatives with respect to capital and operation and maintenance costs, permitting, treatment requirements, and feasibility. Further, the city considered alternatives' benefits in terms of increasing the city's source water supply. This source is constrained by groundwater pumping limits imposed by the city's consumptive use permit. Additionally, this option is consistent with the regional Water Management District's policies promoting the use of alternative water supplies. The city then selected the ZLD process after bench-scale and pilot-testing programs demonstrated its feasibility.

As shown in Figure 8-3, the primary components of the ZLD process include an equalization tank, solids-contact clarifiers (SCCs), chemical storage and feed systems, a UF feed system, a pressurized UF membrane system, a UF backwash/cleaning system, a disinfection system, a process recycle and recovery system, a solids handling system, a sludge dewatering system, process and yard piping, electrical systems, and instrumentation and control systems. Table 8-2 presents the water quality parameters used for the design of the ZLD process.

The ZLD process includes a lime and soda ash softening process for reducing calcium hardness, TOC, and color. Lime and soda ash are used to precipitate calcium hardness at a pH of approximately 11.0 in the SCCs. Approximately 0.4 mg/L of polymer is also added to enhance settling. Sulfuric acid and ortho-polyphosphate are added to softened water in the SCCs' effluent launders to achieve pH stabilization and scaling control, respectively. The pressurized vertical UF system in the ZLD process removes the majority of the remaining suspended solids. The filtrate is then conveyed to the UF filtrate/backwash tanks where it is disinfected with chloramines before blending with the combined LPRO permeate and raw water bypass water directly upstream of the existing ground storage tank.

In addition to the RO concentrate, all other waste streams generated within the ZLD process—including UF membrane backwash, UF feed tank blowdown, strainer backwash, gravity thickener supernatant, and belt filter press filtrate—are recycled to the equalization tank to achieve a targeted recovery of 98.5 perent. The remaining 1.5

Table 8-2 ZLD water quality parameters

Parameter	RO Permeate	RO Concentrate	ZLD Water Goal[1]	Actual ZLD Water Quality[2]
Total hardness (mg/L as $CaCO_3$)	4.1	1,650	<155	140
True color (CU)	0	53	<65	40
Total dissolved solids (mg/L)	54	2,500	<2,320	770
Iron (mg/L)	0.02	1.7	<0.17	0.01
Chloride (mg/L)	11.1	600	<1,368	280
Turbidity (NTU)	0.10	8.8	<0.41	<0.15

[1] Based on technical proposal from original equipment manufacturer
[2] Based on WTP No. 2 sample results on Feb. 16, 2015

Table 8-3 UF membrane design criteria

Parameter	Design Value
Design membrane flux	25.3 gfd
Membrane modules per train	24
Treatment capacity	278 gpm
Filtrate flow per module	10.4 gpm

percent of the concentrate is lost as the moisture content in the dewatered lime softening sludge and via the neutralized, chemically enhanced backwash cleaning water used by the UF system.

The settled lime softening sludge is removed from the bottom of the SCCs and pumped to a gravity sludge thickener. Solids removed from the gravity sludge thickener are then dewatered with belt filter presses after polymer addition. The belt filter presses produce a sludge cake with approximately 65 percent dry solids. The original design dry solids was 50 percent. It was determined during startup that a higher dry solids was easily achieved. Once dewatered, the lime sludge is hauled from WTP No. 2 by the city's contract hauler, mixed with crushed concrete, and used as roadway base material.

UF system overview. The UF system includes four pressurized UF membrane skids with each skid having 24 vertical membrane modules with a combined capacity of 0.4 mgd (278 gpm). The total design capacity of 1.2 mgd (833 gpm) is achieved with three skids. The fourth provides redundancy while another skid is in backwash or cleaning mode. Because of the presence of high TOC and antiscalant concentrations, a membrane flux of 25.3 gfd is used for the UF system operation. Table 8-3 provides a summary of the UF system design criteria.

Operations Experience

Process flow management. One of the biggest challenges for the project was the management of all of the recycle streams to increase the recovery of the ZLD process to a minimum of 98.5 percent. Because of the variation in volume, duration, flow, and frequency of each respective recycle stream, an 80,000-gal equalization tank was provided for flow equalization. These recycled streams are mixed and pumped back to the head of the ZLD process. In addition to providing the requisite storage for the recycle flows, the equalization basin was also fitted with a mixing system that includes jet-type mixing eductors and an associated pumping system to keep solids from settling in the tank. Three recycle pumps were installed in the equalization tank to transfer the respective recycle flows

Table 8-4 Concentrate ZLD design flows

Parameter	Units	Value
ZLD design capacity (LPRO concentrate flow only)	mgd	1.20
SCC design capacity (RO concentrate and recycle flows)	mgd	1.65
Existing WTP raw water bypass design capacity	mgd	1.58
Total design production capacity of the WTP	mgd	7.58

LPRO—low-pressure reverse osmosis, SCC—solids-contact clarifiers, WTP—water treatment plant, ZLD—zero liquid discharge

back to the head of the ZLD process for additional treatment. Even with the relatively large capacity of the equalization tank, flow control of the recycle stream was required to avoid upsets to the treatment process. As a result of the recycle flows, the ZLD process units were designed to treat flows higher than the 1.2 mgd concentrate flow to accommodate the recycle streams. The ZLD process design flows are presented in Table 8-4.

The equalization tank was installed below grade so that the various recycle streams could flow to the tank by gravity. Self-priming pumps were selected for recycle pumping from the equalization tank in large part because of the corrosive nature of the combined waste streams in the tank. Submersible pumps would have needed to be fully constructed of stainless steel to withstand the environment, which was not economically feasible. Recycle stream flow control evaluations considered variable frequency drives, motorized v-port ball valves, and the use of multiple pumps. Variable frequency drives were eliminated from consideration because of concerns regarding maintaining prime of the pumps while motor speeds were lower. Three pumps, along with flow-controlled, v-port ball valves on the common recycle line, were selected for implementation.

Disinfection byproducts. Another challenge of the project was determining how the ZLD-treated water would be disinfected without exceeding disinfection byproduct (DBP) limitations. The original concept was to combine the ZLD-treated water stream with the LPRO permeate and raw water bypass streams in the clearwell. However, higher TOC concentrations in the ZLD-treated water caused a concern about DBP formation potential. Because the combination of lime/soda ash softening and membrane filtration provides only a 2-log inactivation of viruses, the ZLD-treated water was disinfected separately from the LPRO permeate using free chlorine to provide the remaining 2-log virus inactivation required for compliance with the Ground Water Rule before chloramination. After the formation of chloramines, the ZLD-treated and disinfected water is combined with the finished water from the facility. The separate disinfection operation minimizes impact to the existing clearwell and reduces overall DBP formation potential.

Integration with the existing facility. The instrumentation and supervisory control and data acquisition (SCADA) systems associated with the ZLD process were completely integrated into the facility's SCADA system to provide for seamless and automated operations, monitoring, and control. This operational approach was in accordance with the operating permit that was issued to incorporate the ZLD process for the LPRO concentrate and to increase the overall WTP capacity. The electrical system for the ZLD process was also integrated into the existing facility's electrical system and includes emergency power for the overall facility in accordance with regulatory requirements. The design intent for the ZLD process was successfully implemented and provides the operators with the ability to monitor and control the operations of the entire facility from the existing operations room. Additionally, most of the ZLD processes can both be monitored and controlled from the UF building. Figure 8-4 presents one of the SCADA screens that shows the overall WTP.

Cost

The capital cost of the ZLD project was approximately $9.5 million.

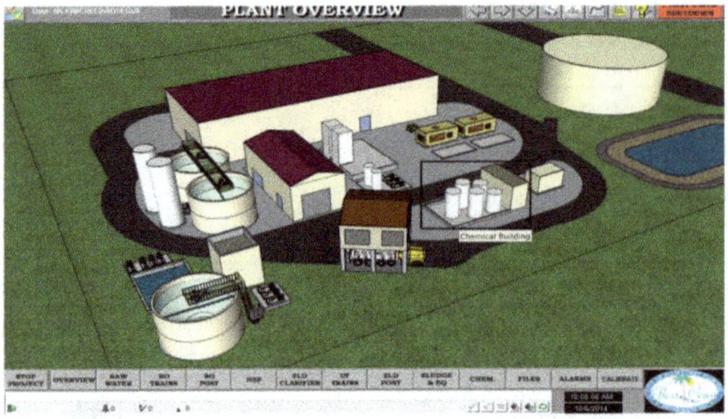

Figure 8-4 Screenshot of ZLD SCADA system

IRWIN WATER WORKS, ARMY NATIONAL TRAINING CENTER, FORT IRWIN, CALIF.

The Fort Irwin National Training Center (Ft. Irwin) is located in the Mojave Desert in California and serves as a major combat training center for the US military. Ft. Irwin uses a groundwater supply that exceeds the maximum contaminant levels for fluoride and arsenic. Water demands average between 2.0 and 2.5 mgd and range from 0.8 to 5.3 mgd, depending on troop rotation and season.

Ft. Irwin was operating two separate water distribution systems. Potable water was produced by treating the groundwater with RO and distributing it through the RO distribution system. Untreated groundwater was distributed through the domestic water distribution system for other uses. However, the California Division of Drinking Water determined that the water delivered via the RO distribution system must meet all applicable drinking water standards. To meet these requirements, Ft. Irwin constructed a new water treatment plant, the Irwin Water Works (IWW), to produce up to 6.0 mgd of potable water, and made improvements to combine the two existing distribution systems to achieve a single potable water distribution system.

Because of high silica concentrations, the RO treatment process operated at a low recovery rate. Pilot testing was performed to select a more efficient treatment process. This resulted in the selection of EDR as the primary treatment process for the new IWW. The EDR process was shown to effectively meet treatment goals while providing a higher recovery rate (92 percent) than RO. The overall recovery rate will be increased to greater than 99 percent by providing additional treatment for the EDR concentrate. The final residuals will be discharged to onsite drying beds, resulting in a ZLD facility. Figure 8-5 depicts the location of the facilities. Figure 8-6 shows the overall process.

Membrane System Overview

The Ft. Irwin EDR system consists of four cartridge filters and five EDR trains. Each EDR train has eight lines of membrane stacks. Each line has three membrane stacks in series. Each membrane stack has 600 cell pairs. A cell pair consists of an anion mebrane, cation membrane, diluate spacer, and concentrate spacer. Table 8-5 presents the EDR system design criteria. Table 8-6 presents the average EDR feedwater, product water, and concentrate water qualities.

EDR Concentrate Management

The facility's overall recovery will be increased to greater than 99 percent by further treating the EDR concentrate. EDR concentrate will be dosed with hydrochloric acid to partially convert bicarbonate to carbon dioxide and then sent through an air stripper to remove

Figure 8-5 Facility location map

alkalinity. The concentrate will then be sent to a lime solids-contact clarifier to remove hardness as well as silica to a target concentration of 12 mg/L. The softened concentrate will be further treated with MF membranes and three-stage RO. The MF system will operate at a recovery of 96.5 percent and a design flux of 24–38 gfd. The RO system will operate at a recovery of 90 percent with design fluxes of 9.7 gfd (stage 1), 9.3 gfd (stage 2), and 8.8 gfd (stage 3). The RO permeate will be returned to the front of the plant where it will blend with untreated groundwater. The RO concentrate will be treated by weak acid cation exchange and sent to a mechanical evaporator (ME). The ME will operate at a recovery of 80 percent. The ME distillate will be returned to the front of the plant. The final ME

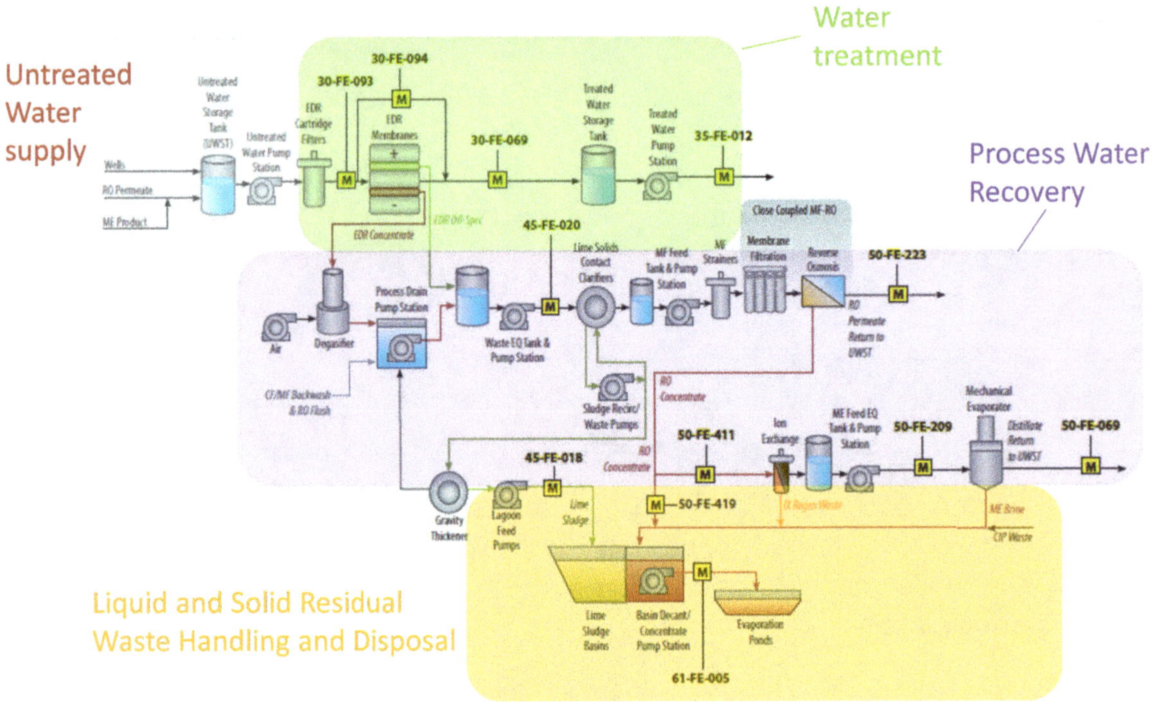

Courtesy of CDM Smith

Figure 8-6 Irwin Water Works process flow diagram

Table 8-5 EDR system design criteria

Item	Value
Number of trains	5
Product flow rate, per train	1.5 mgd
Feed flow rate, per train	1.6 mgd
Number of lines, per train	8
Number of stacks, per train	24
Number of cell pairs, per stack	600
Hydraulic stages, per stack	1
Electrical stages, per stack	1

concentrate will be discharged to onsite evaporation ponds. Thickened lime sludge will be discharged to lime sludge lagoons. The solids in the lime sludge lagoons and evaporation ponds will be periodically trucked off for landfill disposal.

Costs

The IWW project was delivered as a design-build contract. The total cost of the facility and conversion of the dual-piped distribution system to a single potable water distribution system was $100 million. No additional details related to the operation and maintenance cost were available at the time this manual was prepared.

Summary

EDR is a membrane process that can effectively remove arsenic, fluoride, nitrates, and TDS in high fouling waters with minimal pretreatment and a high recovery. The EDR system

Table 8-6 EDR water qualities

Parameters	EDR Feed	EDR Product	EDR Waste
Calcium (mg/L)	24.75	1.8	289
Magnesium (mg/L)	4.2	0.4	48.4
Sodium (mg/L)	167.8	21.9	1,845
Potassium (mg/L)	8.9	0.9	101
Strontium (mg/L)	0.4	0.02	4.78
Barium (mg/L)	0.02	ND	0.23
Arsenic (μg/L)	25	2.7	281
Bicarbonate (mg/L)	178.6	30.3	1,849
Sulfate (mg/L)	123.9	9.5	1,440
Chloride (mg/L)	104.6	11.4	1,182
Fluoride (mg/L)	5.1	0.9	53.5
Nitrate (mg/L)	20.3	2.4	226
Silica (mg/L)	52.3	52.3	52.3
Total dissolved solids (mg/L)	690	132	7,099
Total hardness (mg/L as $CaCO_3$)	79.5	6.0	921
Conductivity (μS/cm)	1,005.8	134	9,089
pH	7.79	7.02	7.85

ND—No data

at IWW will produce 6 mgd of product water at a recovery of 92 percent. The waste from the EDR system will be further treated to increase the overall plant recovery to greater than 99 percent while discharging no liquid waste. The IWW project was delivered as a $100 million design-build contract.

The Ft. Irwin facility provides a critical function in the preparedness of US defense capabilities. A critical factor in sustaining the Ft. Irwin facility is a safe and reliable potable water supply to serve the military personnel, their families, government civilian staff and contractors, and visitors. Using EDR at IWW will allow Ft. Irwin to meet its goals of producing high-quality drinking water while conserving its limited resources to provide a reliable water source for the base and its personnel for many years to come.

FARGO WASTEWATER TREATMENT FACILITY, EFFLUENT REUSE FACILITY

The city of Fargo, N.D., owns and operates a 25-mgd wastewater treatment facility (WWTF) with an average daily flow of 12 mgd. The WWTF treatment processes consist of headworks, screen, trickling filters, denitrification, and chlorine disinfection with effluent discharged to the Red River of the North, a river shared by North Dakota and Minnesota. When the river stage is high, the effluent is pumped from the WWTF to wastewater stabilization ponds where it is stored until it can be discharged to the Red River. Fargo Effluent Reuse Facility (ERF) was built in 2008 to treat approximately 2.1 mgd of the Fargo WWTF effluent and supply the treated water to a 100-million-gpy ethanol plant in Casselton, N.D.

The ERF treatment processes consist of UF membranes followed by RO membranes with a 1.4-mgd RO permeate production capacity. RO permeate is pumped approximately 25 miles from the ERF to the ethanol plant. Cooling tower blowdown from the ethanol plant is returned to the WWTF. UF and RO systems at the ERF were commissioned in October 2008 and started sending RO permeate to the ethanol plant in December 2008. The UF process serves as a pretreatment step for the RO process. Figure 8-7 depicts the location of the facilities. Figure 8-8 shows the overall recycling process.

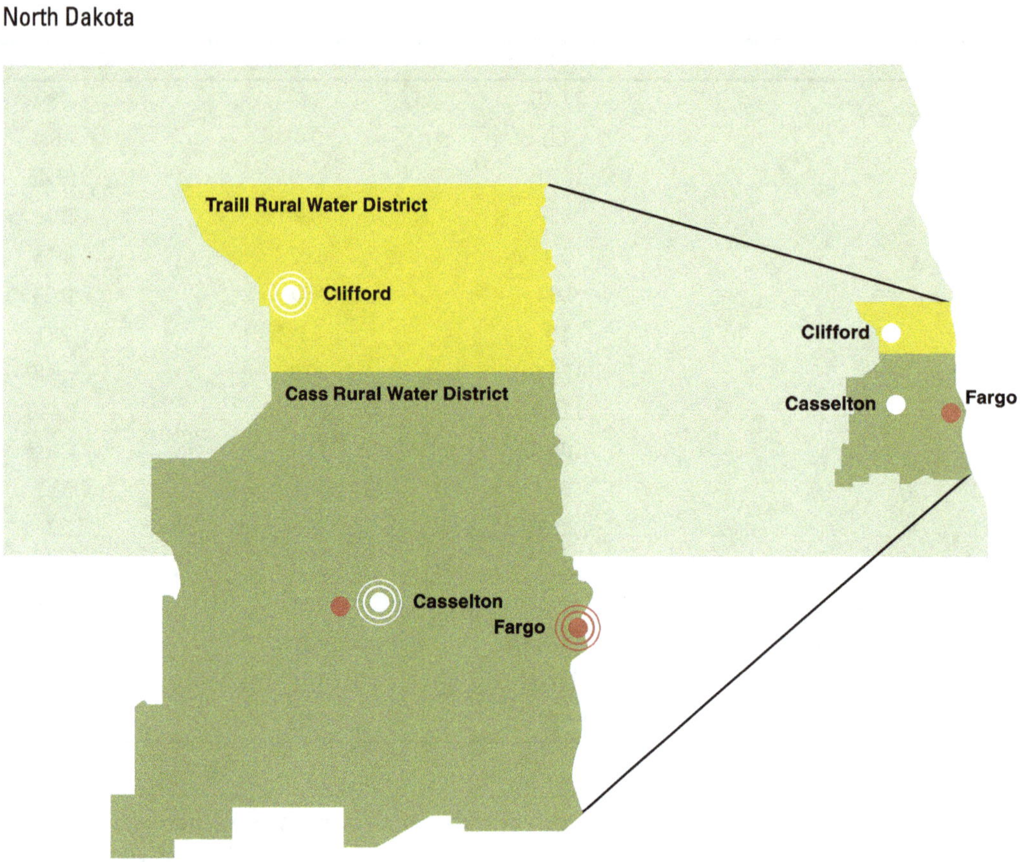

Figure 8-7 Facility location map

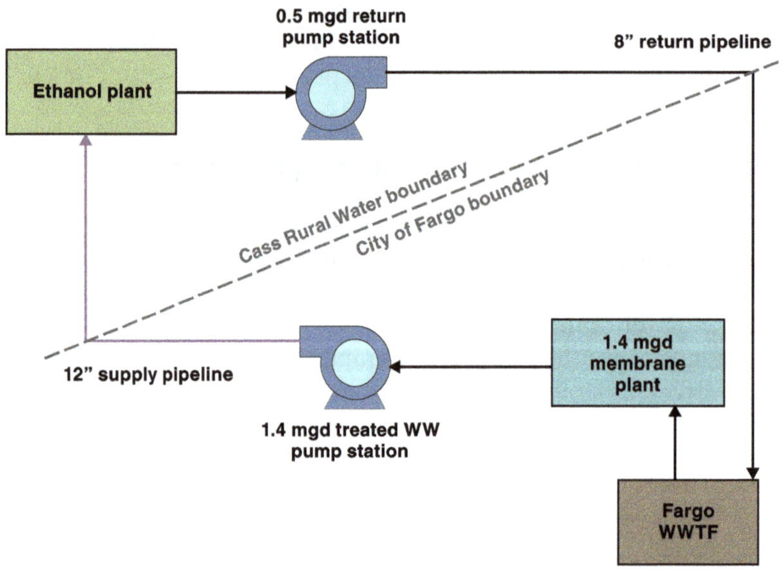

Figure 8-8 Fargo wastewater reuse project schematic

Membrane System Overview

The ERF's UF membrane system consists of two inline 500 μ automatic strainers, three Siemens Memcor CP120 skids, and an associated CIP system. Each UF skid has 120 membrane modules fabricated of polyvinylidene fluoride (PVDF) and operated in an "outside-in" dead-end mode. As shown in Table 8-7, the filter area of a single membrane module is 410 ft^2, and each UF skid provides a total of 49,200 ft^2 active filter area. The design ambient flux is 22 gfd with the maximum allowable transmembrane pressure of 22 psi. Ferric chloride is used as a coagulant and fed to UF influent; after coagulant addition, UF influent passes through two automatic strainers to remove turbidity and organics before reaching the UF membrane. To enhance chemical mixing, a stainless steel static mixer was installed downstream of the coagulant feed addition point. The UF membrane backwash waste is recycled back to the head of the WWTF and membrane cleaning waste is neutralized before discharge to the Red River. Table 8-7 presents the principal UF design criteria.

The ERF's RO system consists of 5 μm cartridge filters, four RO skids, and an associated CIP system. A two-stage RO skid has a 10 × 5 pressure vessel array and six 8-in. elements per pressure vessel. The design membrane flux is 11 gfd. As shown in Table 8-8, each RO skid contains 90 polyamide spiral-wound RO elements, which provide a total of 36,900 ft^2 filter area. The pressure vessel is designed for continuous use at any positive pressure up to 600 psi at 120°F. The overall recovery of the RO system has been running at 70 percent in the past five years, with each skid producing 250 gpm RO permeate. The

Table 8-7 UF membrane characteristics

Item	Value
Number of skids	3
Membrane module	120 per skid
Materials	Polyvinylidene fluoride (PVDF) hollow fiber
Filter area	410 ft^2 per module
Total filter area	49,200 ft^2 per skid
Operation temperature	<45°C
pH	2–10
Nominal membrane pore size	0.04 μm
Design flux	22 gfd

Table 8-8 RO unit parameters

Item	Value
Number of RO skids	4
Membrane module	6 per vessel, 90 per skid
Materials	Polyamide (thin composite layer)
Module area	410 ft^2 per module
Design flux	11 gfd
Recovery	70%
CIP interval	6 months

CIP—clean-in-place, RO—reverse osmosis

Table 8-9 ERF water quality

Parameters	UF Feed	RO Feed
Chloride (mg/L)	114	119
Calcium (mg/L)	76.3	74.6
Silicon (mg/L)	6.44	6.54
Biochemical oxygen demand $(BOD)_5$ (mg/L)	7.2	<2.0
pH	7.65	7.93
TDS (mg/L)	977	968
Total suspended solids (mg/L)	11.8	<1.0
Turbidity (NTU)	7.21	0.295
Conductivity (μS/cm)	1,456	1,491
TOC (mg/L)	10.8	9.0
Magnesium (mg/L)	49.6	48.9
Sodium (mg/L)	138	137
Potassium (mg/L)	17.0	16.6
Iron (mg/L)	0.417	0.044
Manganese (mg/L)	<0.020	<0.020
Sulfate (mg/L)	340	301
Nitrate – N (mg/L)	21.9	22.0
Ammonia as N (mg/L)	0.16	1.54
Total phosphorus as PO_4 (mg/L)	8.95	8.15
Alkalinity as bicarbonate (mg/L)	154	156

TDS—total dissolved solids, TOC—total organic carbon

average RO concentrate flow is approximately 428 gpm. Table 8-8 presents the principal RO process design criteria. Table 8-9 depicts UF and RO feedwater quality.

RO Concentrate Management and Permit

Surface water discharge, sanitary sewer disposal, land application, evaporation ponds, and deep-well injection were evaluated for the ERF's RO concentrate discharge. Surface discharge to the Red River was identified as the most realistic and practical option considering the social, environmental, and economic impact. Because the WWTF already had a North Dakota Pollutant Discharge Elimination System (NDPDES) permit that allows the facility to discharge the secondary effluent to the Red River, the owner decided to approach the North Dakota Department of Health (NDDH) about the possibility of blending RO concentrate with the WWTF secondary effluent and discharge blending to the Red River under the same permit. Extensive model simulations were performed to assess the impact of commingling these discharges, and the NDDH ultimately approved this approach. The blended stream is required to meet all discharge regulations delineated by the NDDH. At the time of any operational challenges and/or upset conditions, the RO concentrate will be sent back to the headworks of the WWTF and commingled with the WWTF influent for treatment.

Operations Experience

In the past few years of operation, foaming issues were identified in the mixing basin where the RO concentrate blends with the WWTF's secondary effluent. The foaming became a public nuisance to the residents in the vicinity, which eventually prompted the Fargo ERF to reroute the RO concentrate to the WWTF's headworks and to subsequently go through the treatment process to mitigate the foaming concern.

Costs

The operational cost associated with this option to discharge RO concentrate to the Red River was low because no incremental pumping or chemicals were required by this approach to disposal. When the RO concentrate is returned to the head of the Fargo WWTF, the cost associated with the RO concentrate management is equal to the average cost to treat the municipal wastewater. No additional details related to the operation and maintenance cost were available at the time this manual was prepared.

KAY BAILEY HUTCHISON DESALINATION PLANT, EL PASO, TEX.

El Paso Water operates the 27.5-mgd Kay Bailey Hutchison brackish groundwater desalination plant (KBH), which at the time of this writing is the largest inland brackish groundwater RO plant in the United States. As Figure 8-9 shows, the KBH is located near El Paso, Tex., on Ft. Bliss property, east of the El Paso International Airport. The plant was brought online in August 2007 to provide reliable water supply for Ft. Bliss and the City of El Paso. It is named after former Sen. Kay Bailey Hutchison, who helped facilitate the public–public partnership between Ft. Bliss and the City of El Paso.

As Figure 8-10 shows, the KBH treatment processes consist of pretreatment, membrane desalination, and permeate treatment. Sixteen wells provide brackish

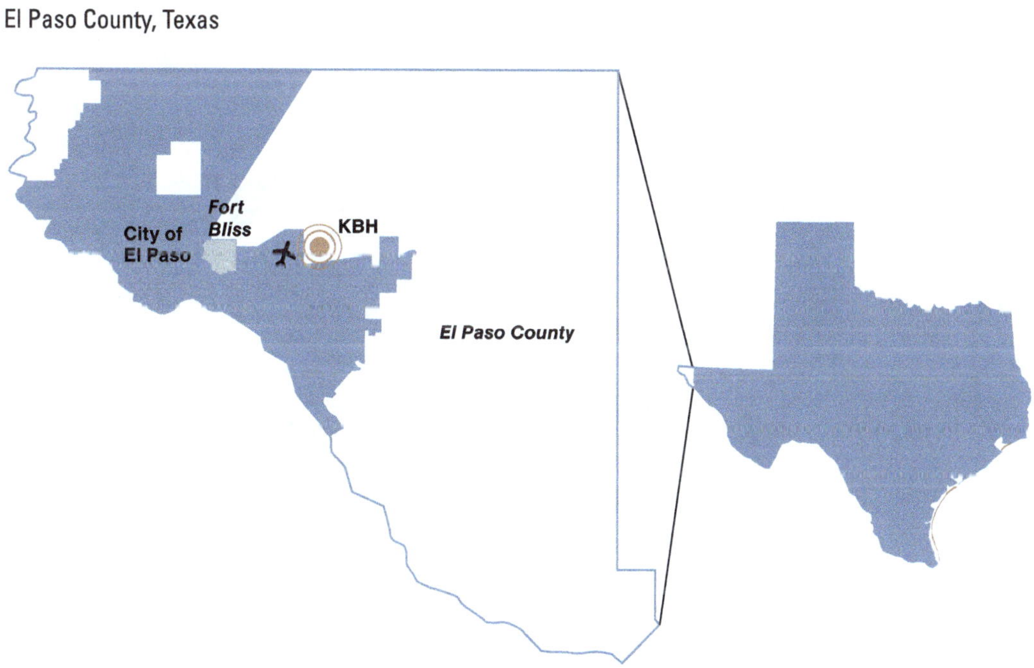

Figure 8-9 Kay Bailey Hutchison Desalination Plant location (El Paso County, Tex.)

groundwater as feedwater, which is pretreated by screening through 25-µm stainless steel mesh screens; dosing with antiscalant (and possibly acid, though not currently); and filtering through 15-µm cartridge filters. The KBH plant has five trains of two-stage RO, each with a design permeate flow of 3 mgd for a total permeate flow of 15 mgd. RO concentrate is pumped approximately 22 mi from the KBH plant for concentrate disposal by gravity-driven deep-well injection. An additional 16 wells provide water for blending with the RO permeate, which provides the plant with a total capacity of 27.5 mgd. Blended effluent is dosed with hypochlorite and polyorthophosphate before distribution.

Membrane System Overview

The KBH desalination system (Figure 8-11) consists of five two-stage RO trains configured in a 2:1 array and a CIP system. As Table 8-10 shows, each RO train has seven 8-in. Hydranautics ESPA1 elements per pressure vessel. Consequently, each train has 201,600 ft^2

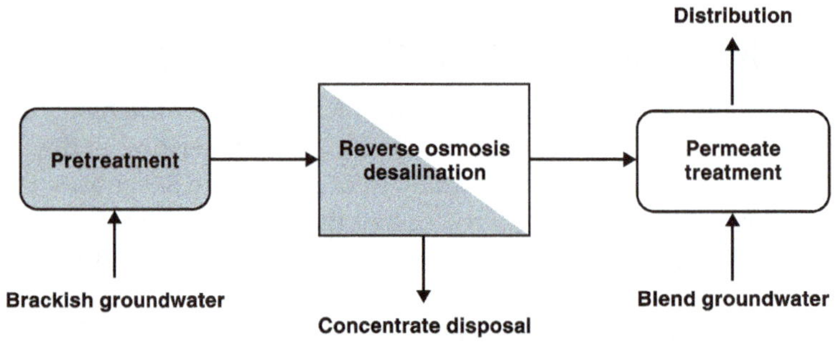

Figure 8-10 Kay Bailey Hutchison plant desalination process schematic

Courtesy of El Paso Water

Figure 8-11 Kay Bailey Hutchison plant RO trains

Table 8-10 RO unit parameters

Item	Value
Number of RO trains	5
Train array	8-in. pressure vessels, 7 elements per vessel; 1st stage – 48 vessels per train; 2nd stage – 24 vessels per train
Membrane elements	Hydranautics ESPA1 8 in. × 40 in.
Membrane area	400 ft^2 per element; 201,600 ft^2 per train
Permeate	3.0 mgd (2,083 gpm) per train
Average flux	14.9 gfd
Flux balancing	Permeate throttling
Overall recovery	83%
CIP Interval	12 months

CIP—clean-in-place, RO—reverse osmosis

Table 8-11 KBH water quality (February 2015)

Parameter	RO Feed	RO Permeate	RO Concentrate
pH	7.6–7.8	6.4–6.6	7.9
Conductivity (μS/cm)	4,200–5,000	550–900	21,000–26,000
TDS (mg/L)	2,500–3,000	270–420	12,000–16,000
Calcium (Ca^{2+}, mg/L)	130–160	2–3	650–910
Magnesium (Mg^{2+}, mg/L)	30–50	0.4–0.5	160–320
Potassium (K^+, mg/L)	16–22	2.4–3.2	78–91
Sodium (Na^+, mg/L)	770–880	105–165	3,150–3,960
Alkalinity (mg/L as $CaCO_3$)	80–85	<20	405–420
Chloride (Cl^-, mg/L)	1,180–1,450	170–260	5,500–7,000
Sulfate (SO_4^{2-}, mg/L)	230–350	<5	1,300–1,950
Silica (SiO_2, mg/L)	31–38	5.0–5.5	140–160

RO—reverse osmosis

of membrane area with an average membrane flux of 14.9 gfd. The system typically operates at 83–84 percent recovery with a first-stage feed pressure of 170–185 psi. Flux balancing between the first and second stage is controlled by throttling the first-stage permeate with a backpressure of 30–40 psi.

The brackish feed salinity varies 2,500–3,500 mg/L depending on which brackish wells are being pumped and how long they have been pumped. Permeate TDS varies from 250 to 450 mg/L; the permeate is blended with fresh groundwater to produce a finished water TDS of approximately 400–600 mg/L. Table 8-11 lists the typical water quality parameters of the KBH plant for February 2015.

RO Concentrate Management and Permit

El Paso Water has a Texas Commission on Environmental Quality (TCEQ) Underground Injection Control (UIC) Class V authorization that allows the KBH plant to discharge up to 3,363 acre-ft per year (3 mgd) of RO concentrate to a deep-well injection site with three wells (approximately 22 mi northeast of the KBH plant). El Paso Water obtained approval from USEPA and TCEQ for an exempt aquifer permit, which is a waiver of the requirement that the injected concentrate meet primary drinking water standards. The TCEQ Class V authorization has been amended to reflect the waiver of the water quality requirement for the injected concentrate. No adverse effects have been observed on the water level or quality of the receiving aquifer.

Operations Experience

When the KBH plant started up in August 2007, the feed salinity was 2,000–2,500 mg/L, but over the eight years of operations, the feed salinity has increased to 2,500–3,500 mg/L. The operators currently rotate which wells are pumped to try to maintain low feed salinity.

During peak demand seasons (summer months), the plant is operated at full capacity, but production is cut back during nonpeak seasons. During winter months, the plant typically operates only one train (out of the five) at a time. The plant alternates trains on a weekly basis and performs permeate flushes three times per day on the nonoperational trains.

Originally, the KBH plant was designed with 5-μm cartridge filters, but the fouling and replacement of those cartridge filters required excessive labor and cost. So the cartridge filters were replaced with 15-μm cartridge filters, which has resulted in significantly less labor and lower maintenance expenses with no adverse effects on the RO performance.

Costs

Initial capital costs for the KBH project were $91 million (2005), which included approximately $32 million for production wells and collectors; $40 million for the desalination plant and near-plant piping; and $19 million for concentrate disposal. As of 2014, the combined capital and operating costs were estimated to be $1.99/kgal of product water ($648/acre-ft).

Public–Private Partnership for Mineral Recovery From Concentrate

The El Paso City Council approved incentives for Enviro Water Mineral Co. (EWM) to construct a $65M industrial facility adjacent to the KBH plant to recover saleable minerals from the KBH concentrate[1]. The EWM facility had its ribbon-cutting event on April 25, 2017, and employs treatment processes such as softening, NF, and ED to recover chemicals (caustic and hydrochloric acid) and minerals (gypsum and magnesium hydroxide). The treatment process is expected to recover up to 1.0 mgd of water from the concentrate, which will be returned to El Paso Water for drinking water production. This partnership is expected to increase the recovery of the system from 83 to 99 percent. The EWM process provides for additional concentrate disposal capacity, similar to that of adding another injection well. Added concentrate disposal capacity allows the KBH to treat more brackish groundwater, which will allow for increased production from the plant.

Acknowledgments

Special thanks to El Paso Water: Art Ruiz, Scott Reinert, Fernie Rico, Gilbert Trejo, and John Balliew.

[1] Martinez, A. 2015. Council OKs Deal for $65M Water Plant. *El Paso Times*, Oct. 13.

ZERO LIQUID DISCHARGE WATER TREATMENT FACILITY

To comply with State of California requirements for drinking water quality and wastewater effluent discharge, a California agency embarked on a facility enhancement project to improve both the water and wastewater quality. A 2003 water system upgrade study determined that the existing groundwater supply was the only option available and that RO treatment was the most appropriate technology for meeting the water quality requirements. Providing drinking water with lower TDS also solved a wastewater effluent discharge issue. At this inland location, ZLD was the only alternative for dealing with the RO concentrate, and a combination of a mechanical evaporator and solar evaporation ponds was selected as the treatment approach. The evaporator is a falling film vapor compression brine concentrator, which heats approximately 154 gpm of RO concentrate as it enters the evaporator. The facility recovers more than 99 percent of the water as drinking water by combining the highly pure distillate water vapor with the RO permeate. Approximately 4 gpm is discharged to the double-lined evaporation ponds that required special permitting.

Treatment System Overview

Figure 8-12 presents an overview of the treatment system. Approximately 0.8 mgd of brackish groundwater from four onsite wells is conveyed to the RO treatment step. Pretreament for the RO process includes a three-step filtration process designed to remove sand and protect the RO system from particulates. Chemicals (i.e., acid, scale inhibitor, and sodium bisulfite) are added to condition the raw water. The acid addition allows for control of iron and manganese and for alkalinity recovery in the finished water. Sodium bisulfite is added to scavenge dissolved oxygen to further prevent iron and manganese precipitation and to reduce corrosion. The RO system itself is a two-stage design with a recovery of 70–80 percent; Table 8-12 presents some key characteristics.

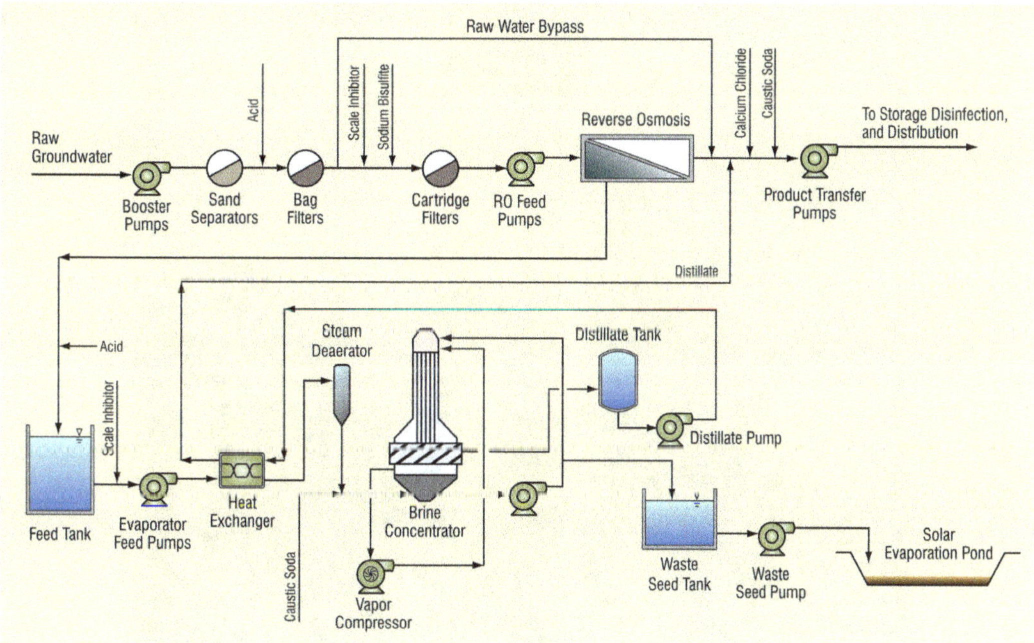

Figure 8-12 ZLD treatment schematic

Concentrate from the RO trains is transferred to an evaporator feed tank and then to a two-step evaporation process that combines mechanical and solar evaporation to dispose of the concentrate. A falling film vapor compression brine concentrator (see Figure 8-13) preheats the RO concentrate, which rapidly evaporates when it enters the unit. The highly pure water vapor is condensed, cooled, and then blended with the RO

Table 8-12 RO system characteristics

Item	Value
Number of trains	2
Number of stages per train	2
RO feed pumps	2 @ 280 gpm and 358 ft total dynamic head
Permeate flow rate per unit	203 gpm
Concentrate flow rate per unit	77 gpm

RO—reverse osmosis

Courtesy of Carollo Engineers, Inc.

Figure 8-13 Falling film vapor compression evaporator

permeate to increase the recovery of potable water from the project. The brine concentrator operates at a recovery of 97.4 percent, recovering all but 4 gpm of concentrate blowdown that is pumped to the solar evaporation ponds for final disposal (Table 8-13). Four lined evaporation ponds were provided. More than 99 percent of the groundwater is recovered as potable water. The ponds also provide storage capacity for the solids in the concentrate from the brine concentrator. No information about the solids storage capacity of the ponds was available to the chapter authors at the time this manual was published.

Because of the low salinity of the RO permeate and the distillate from the brine concentrator, the blended product water stream must be conditioned to stabilize it to protect the water distribution system. This conditioning step is achieved by adding caustic soda and calcium chloride to increase the pH and add some alkalinity and calcium hardness to the water.

Concentrate Management and Permit

As mentioned previously, the final disposal method for the blowdown stream from the thermal evaporator is evaporation ponds. Details of the ponds are presented in Table 8-14, and Figure 8-14 is a photograph of one of the ponds.

The solar evaporation ponds are triple-lined and are classified as a Class II Surface Impoundment in California by the Regional Water Quality Control Board (RWQCB), requiring special monitoring and reporting. To meet the requirements, the ponds are designed to contain the 1,000-year storm event and also withstand the maximum credible earthquake. A major consideration for the ponds is that they have a zero-leakage requirement, which can be confirmed by the leachate collection and monitoring system that is installed. Regular reporting to the RWQCB is required to confirm the integrity of the ponds.

Table 8-13 Brine concentrator system

Item	Value
Number of units	1
Type of unit	Falling film vapor compression
Operating temperature	~220°F
RO concentrate influent flow rate	154 gpm
Distillate	150 gpm
Concentrate	4 gpm
Concentrate TDS	Up to 250,000 mg/L
Concentrate solids concentration	5–7%

RO—reverse osmosis, TDS—total dissolved solids

Table 8-14 Evaporation ponds

Item	Description
Number of ponds	4
Type of ponds	Lined solar evaporation
Integrity monitoring	Leachate collection with flow sensors
Pond area	~4 acres
Average evaporation rate	1 gpm per acre

Courtesy of Carollo Engineers, Inc.

Figure 8-14 One of the triple-lined evaporation ponds used for final disposal of concentrate from the thermal evaporator

Operations Experience

The water treatment system and brine disposal facilities were placed into operation in 2006. There were some initial challenges with optimizing the performance of the vapor compression evaporator, but since those were sorted out, the plant has been operational. As mentioned, regular reporting to the RWQCB is required to confirm the integrity of the solar evaporation ponds, which adds a layer of operational requirement to the facility. The final blowdown stream from the evaporator contains solids (precipitate) formed during the evaporation process. To avoid causing pipe blockages, the solids are pumped continuously and then discharged to the particular pond that is in service.

Costs

The capital cost of the entire installation was more than $30 million (2006), which includes the RO groundwater treatment system, the evaporator, and the final disposal ponds. In terms of operational costs, the energy costs for the evaporator are high but justified because there is no way other than evaporation to dispose of the concentrate stream at this inland location. Without the evaporator, given that the average evaporation rate is 1 gpm per acre, a very large area for the solar evaporation ponds would be required. No additional details related to the operation and maintenance cost, such as periodic costs, if any, for solids removal from the evaporation ponds (if required), were available at the time this manual was prepared.

LAGUNA COUNTY SANITATION DISTRICT DEEP-WELL CONCENTRATE DISPOSAL

The Laguna County Sanitation District (LCSD) is located in northern Santa Barbara County, Calif. LCSD owns and operates a wastewater treatment plant that serves the unincorporated community of Orcutt and areas in unincorporated Santa Maria as well as portions located within the City of Santa Maria (see Figure 8-15).

The plant, currently treating about 2.1 mgd, comprises two parallel trains; a low-TDS plant where the bulk of the flow is treated, and a high-TDS plant where about 17–25 percent of the influent flow is treated and desalinated using RO. Flow is diverted to the high-TDS plant for approximately 5 hr each day. This corresponds to the time when self-regenerating

Figure 8-15 Facility location map

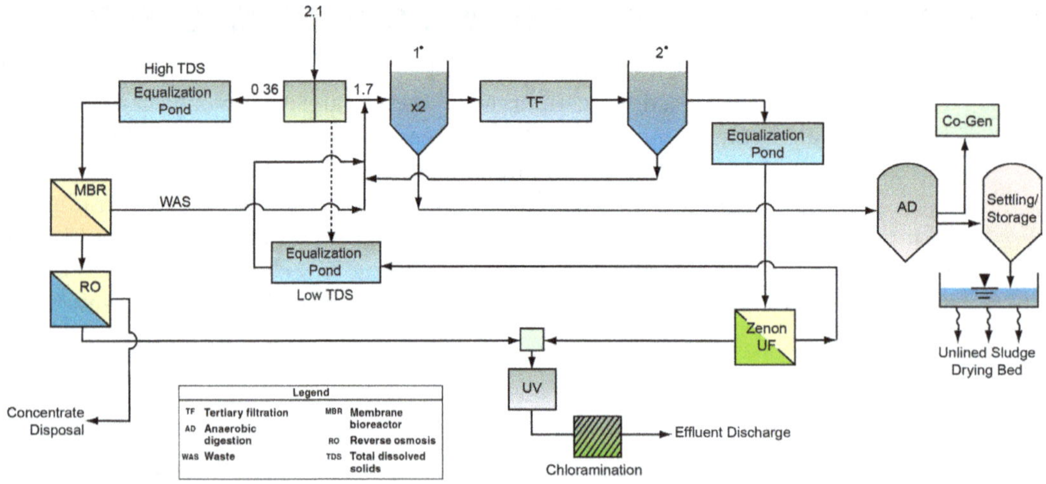

Figure 8-16 LCSD treatment schematic

water softeners in the service area are typically discharging salt to the collection system. The need for segregating and treating the higher-TDS portion of the flow is based on RWQCB Basin Plan objectives for sodium, chloride, and TDS. These objectives are stipulated in the plant's Waste Discharge Requirements (WDRs). TDS, for example, must be less than 900 mg/L to be land applied for discharge.

Treatment System Overview

Figure 8-16 presents an overview of the treatment system at the LCSD plant. As mentioned, the plant receives approximately 2.1 mgd of mostly domestic-type wastewater. After initial screening and grit removal at the plant headworks, the higher-TDS fraction of the influent flow is diverted using conductivity measurement to an equalization (EQ) pond. On the way to the EQ pond, the water passes through fine screens (1 mm). From the EQ pond, the screened wastewater is processed in a membrane bioreactor (MBR) system, which removes organic material (represented by biochemical oxygen demand) and provides some oxidization of ammonia to nitrate. The MBR uses UF membranes that essentially remove all suspended solids from the water, providing good pretreatment and a low-turbidity feedwater for the RO treatment step.

Pretreatment for the RO process also includes chemical addition (acid and scale inhibitor) to condition the raw water for treatment. Key system characteristics for the MBR and RO systems are presented in Tables 8-15 and 8-16.

Concentrate from the RO trains is pumped, using the residual concentrate pressure, through roughly 6 mi of pipeline to an existing well for disposal by deep-well injection. The concentrate is disposed into a Class 1 nonhazardous injection well called Union Sugar Well No. 13, which was originally an oil-producing well.

The plant is regulated by the RWQCB–Central Coast Region under WDRs and Master Reclamation Permit Order R3-2011-0217, adopted Dec. 1, 2011. The injection well is regulated by USEPA permit No. CA 10910004. The most recent permit was issued in March 2010 and is valid for 10 years.

The WDR permit lists various salts in addition to TDS (see Table 8-17).

Concentrate Management and Permit

As mentioned previously, the final disposal method for the concentration from the RO trains is deep-well injection into a Class I nonhazardous injection well. The injection zone

Table 8-15 MBR system characteristics

Item	Value
MBR capacity	0.5 mgd
Number of trains	2
Membrane type and manufacturer	UF/GE ZENON
Capacity per train	0.25 mgd

MBR—membrane bioreactor

Table 8-16 RO system characteristics

Item	Value
RO capacity	0.54 mgd[1]
Number of trains	3 (2 duty)[2]
Number of stages per train	4 (4 × 2 × 1 × 1 array)
Recovery	81%
Permeate flow rate per unit	100 gpm
Concentrate flow rate per unit	24 gpm

RO—reverse osmosis
[1] Capacity is with all trains in operation.
[2] A fourth train is connected, but it currently does not have membranes installed.

Table 8-17 WDR permit limits for discharge

Item	12-month Running Mean Value (mg/L)
TDS	900
Sodium	180
Chloride	180
Sulfate	300
Boron	0.5

TDS—total dissolved solids, WDR—waste discharge requirement

is the Monterey Formation, a fractured shale, at a depth of approximately 4,800–5,300 ft below ground surface. In this case, no pumping of the concentrate down the well is required as it essentially discharges by gravity into the well. The injection formation is confined by the Sisquoc Formation, located at about 4,300–4,800 ft below ground surface.

The district is required to maintain good operation of the well by monitoring and recording certain parameters, as shown in Table 8-18. The well is permitted for a maximum discharge flow rate of 225,000 gpd. Reports are required quarterly by the USEPA.

Operations Experience

The well was placed into service in May 2003. Since that time, the district has performed annual integrity tests as well as periodic cleaning operations to restore the permeability to the original rates, but there is no evidence of loss of capacity. The quarterly reporting to the USEPA has been relatively trouble free. As far as the district is aware, this well is the only Class I well in the state that does not require pressurization for operation.

Table 8-18 Monitoring requirements for deep well

Item	Requirements
Injection rate (gpm)	Continuously/weekly
Daily injection volume (gal)	Weekly
Total cumulative volume (gal)	Weekly
Well head injection pressure (psig)	Continuously/weekly
Annual pressure (psig)	Continuously/weekly
Injection fluid temperature (°F)	Continuously/weekly

Costs

The cost to convert the original oil well to a disposal well was about $1.4 million, with an additional initial well test that was around $180,000. Annual costs for permit compliance and testing are about $35,000, plus small costs for a monitoring and alarm system. Renewal of the discharge permit through the USEPA every 10 years has cost approximately $50,000. The cost to transfer the RO concentrate to the well is incorporated into the RO operating cost.

CHINO DESALTER AUTHORITY CONCENTRATE REDUCTION FACILITY

The Chino Groundwater Basin provides a critical water supply for agriculture, industry, and public drinking water suppliers in the Inland Empire of Southern California. The basin covers more than 220 mi^2. With a safe yield of more than 140,000 acre-ft per year, the basin provides water for more than 700,000 residents. The basin is strategically located amid one of the population centers and major growth areas of Southern California; it is mostly within San Bernardino County but also extends into areas of Los Angeles and Riverside counties (see Figure 8-17).

The basin is an adjudicated basin; as such, the Chino Basin Watermaster was appointed to administer and enforce the provisions of the Basin Judgment. The Chino Basin Desalter Authority (CDA) is a joint powers authority established to operate and manage two groundwater desalters that draw water from the basin and produce drinking water (the Chino I desalter, in operation since 2000, and the Chino II desalter, in operation since 2006).

To recover additional drinking water, the CDA undertook a project to treat the concentrate from the Chino II desalter. Chino II had an initial capacity of 10 mgd and treated flow from eight wells in both RO and ion exchange (IX) processes to remove salinity and nitrate in the groundwater. The capacity of the plant was recently increased to 16.5 mgd. IX regeneration waste and RO concentrate are discharged to the Inland Empire Brine Line for eventual final disposal in the Pacific Ocean. Following pilot testing, the CDA decided to move forward with an enhanced recovery process consisting of a pelletized softening step, followed by filtration, and then a secondary RO process. The Concentrate Reduction Facility (CRF) commenced operation in May 2017. The overall recovery of the Chino II desalter is designed to increase from 85 to 96 percent.

Figure 8-18 shows a schematic of the CRF.

Concentrate Reduction Facility Overview

The Chino II CRF receives flow from the primary RO treatment trains. There are five two-stage primary RO trains that operate at a recovery of 83.5 percent, which is limited by

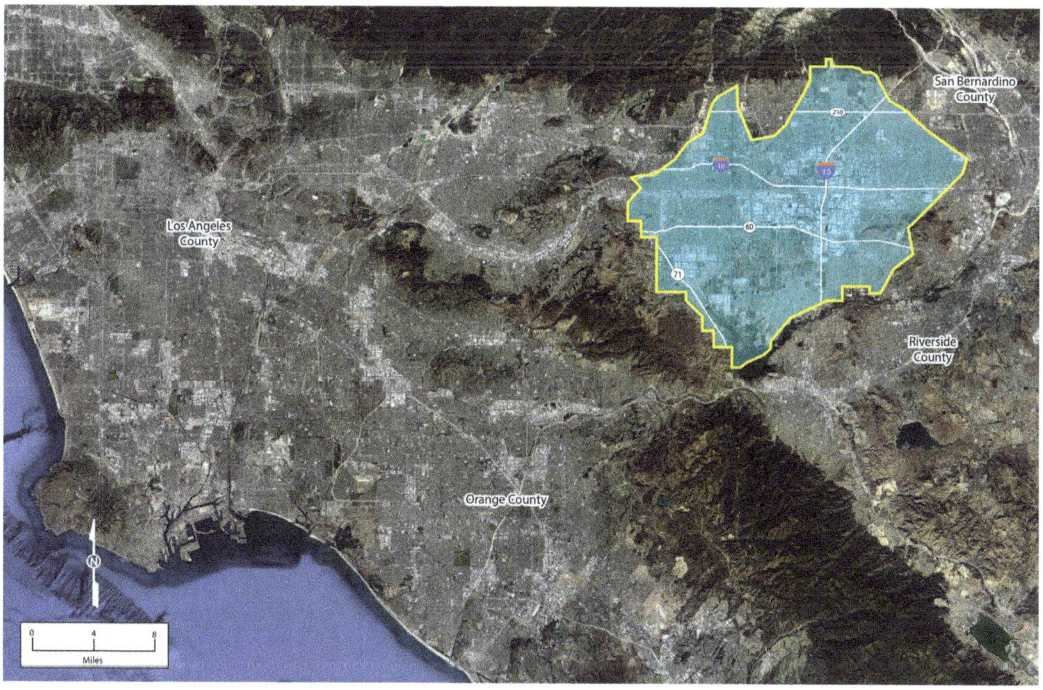

Figure 8-17 Facility location map

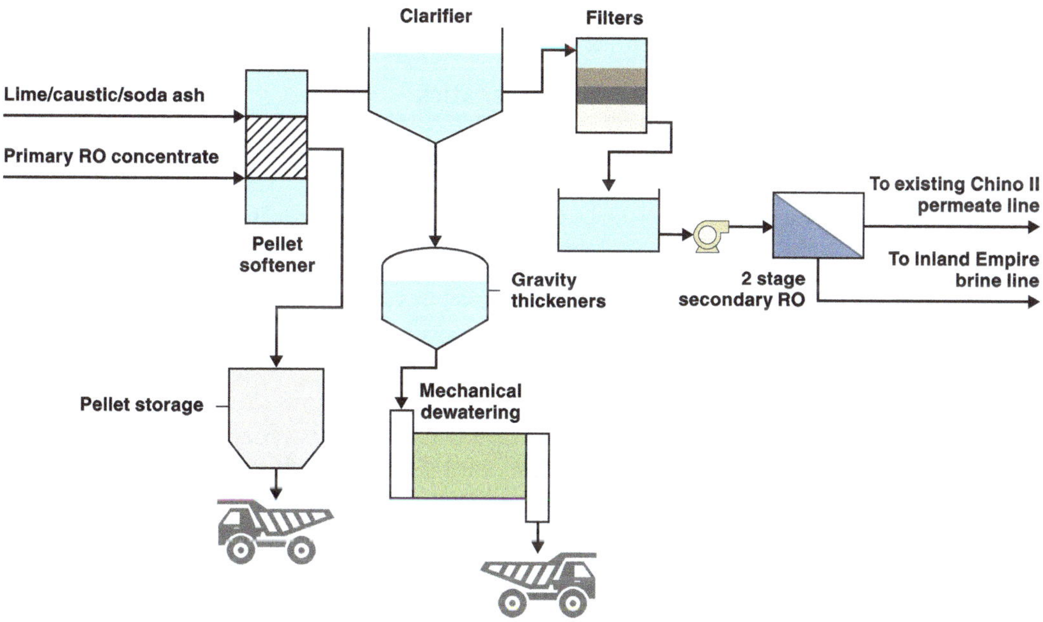

Figure 8-18 Chino II CRF process schematic

the silica concentration of the groundwater. Table 8-19 presents some of the primary RO characteristics.

The CRF receives up to 2.7 mgd of primary RO concentrate and uses chemical softening to remove the limiting foulants in the primary RO concentrate (specifically calcium and silica). Table 8-20 presents some of the major water quality parameters in the primary RO concentrate and the softened water that becomes the feedstream for the secondary RO process. This data is based on the pilot testing, which demonstrated very

Table 8-19 Primary RO characteristics

Item	Value
Number of trains	5
Capacity per train	3 trains of 3.25 mgd; 2 trains of 2.5 mgd
Overall recovery	83.5%
Concentrate flow	1.4–2.7 mgd (depending on conditions)

RO—reverse osmosis

Table 8-20 Water quality summary

Water Quality Parameter	Primary RO Concentrate	Softened Concentrate (Secondary RO Feed)
pH	7.61	8–9.5
TDS (mg/L)	3,319	2,718
Calcium (mg/L)	679	<12
Magnesium (mg/L)	102	<34
Sodium (mg/L)	252	778
Chloride (mg/L)	455	468
Nitrate (mg/L)	524	524
Alkalinity (mg/L as $CaCO_3$)	1,145	302
Silica (mg/L)	194.3	<75

Table 8-21 Secondary RO characteristics

Item	Value
Number of RO trains	3
Capacity per train	365–810 gpm (as permeate)
Overall recovery	66%–85%
Concentrate flow	0.48–0.93

RO—reverse osmosis

good removal of calcium, magnesium, and silica through the softening process, followed by clarification and filtration.

The CRF consists of an energy recovery system upstream of the softening step to recover excess energy from the primary RO concentrate. This is followed by four pellet reactor softeners operating in parallel to remove calcium carbonate and some silica from the concentrate. Lime, caustic soda, and soda ash is used to raise the pH and cause precipitation of the dissolved solids within the pellet reactor. Two solids-contact clarifiers follow to remove carryover calcium, magnesium, and silica particles; the clarifiers are followed by four granular media filters, operating in parallel, to remove fine particles and polish the softened water. Ferric chloride is added to the clarifier feed to assist with settling. The filtered water is fed to three parallel secondary RO treatment trains (Table 8-21). The permeate from these trains is blended with the permeate from the primary RO trains. The concentrate is discharged to the Inland Empire Brine Line.

The pellet softener is seeded with silica sand or calcium carbonate, on which precipitated material forms. Excess pellets from the reactors are blown down regularly to a pellet storage facility and then transported by truck from the site to be reused by local industries. Solids from the contact clarifiers are thickened and then dewatered.

RO Concentrate Management and Permit

Currently, the Chino II Desalter is permitted to discharge up to 1.62 mgd of waste brine into the Inland Empire Brine Line. Under the current mode of operation, the brine flow resulting from the expanded primary RO system at the plant would be closer to 2.7 mgd, including some flow from the IX system, which would be greater than the permit amount. With concentrate treatment of the primary RO brine, the resulting flow to the Brine Line is approximately 0.95 mgd. Concentrate flows from the CRF contribute 0.74 mgd of this total discharge volume, which brings the facility within its permitted discharge flow rate.

Operations Experience

The CRF commenced operation in May 2017 (Figure 8-19). Preliminary operations data demonstrated that its performance meets or exceeds the design targets. Pending publication of operations data, Table 8-22 summarizes the anticipated chemical usage from the pilot testing work.

Courtesy of Carollo Engineers, Inc.

Figure 8-19 Pellet reactors (left) and secondary RO (right) at the Chino II CRF

Table 8-22 Anticipated chemical usage

Chemical	Expected Dose (mg/L)
Lime	200
Caustic soda	820
Soda ash	125
Ferric chloride	20

Costs

The project was bid in 2014, and the winning bid was $47 million. This project produces between approximately 1.0 and 1.9 mgd of additional potable water, which generates additional revenue for the CDA. In addition, the equivalent reduction in concentrate flow to the Inland Empire Brine Line reduced the operational cost associated with the final concentrate disposal step. The addition of the CRF to the CDA's overall water production system resulted in an increase in operation and maintenance costs of less than 10 percent.

Chapter 9

Technological Trends, Salt Recovery, and Beneficial Uses

Ali Sharbat, Charlie (Qun) He, and Lucas DeBuren

This chapter presents a brief synopsis of emerging concentrate treatment alternatives, some of which provide opportunities for salt recovery, nutrient recovery, and beneficial use of concentrate. Unlike the proven mainstream technologies discussed in Chapters 3 through 5, technologies and alternatives included in this chapter were still in the development stage at the time this manual was written. Therefore, for near-term projects, it is recommended that utilities focus on proven technologies to establish a conservative capital improvement budget. However, considering the complexity and fiscal impact of desalination and concentrate management facilities, early proactive planning is recommended. In this case, utilities should not overlook the emerging concentrate treatment alternatives and technologies discussed in this chapter, especially those that are suitable for the site-specific salinity challenge and could mature in the near future (within three to five years).

Moreover, incremental improvements in design, operation, and maintenance of the mainstream desalination technologies (e.g., RO and EDR) are also discussed in this chapter. Some of these technological improvements are continuously developing and could provide additional benefits when being considered during the detailed design phase of the project.

INDUSTRY TRENDS

Hardly a month goes by without hearing reports on new technologies and innovations relevant to brackish water desalination and concentrate management. Most of these innovations can be categorized into one of the following industry trends:

- innovations in membrane technology,
- innovations in design and operation,
- innovative processes,
- improvements in energy efficiency, and
- improved salt recovery and beneficial uses of concentrate.

INNOVATIONS IN MEMBRANE TECHNOLOGY

Through continual improvements, membranes are significantly more efficient and reliable than they used to be. Innovative membrane products, both for NF/RO and ED/EDR as well as other emerging membranes, continue to emerge and offer the promise of better chlorine resistance, lower energy consumption, higher flux, better product water quality, selective removal of target ions, and lower fouling and scaling propensity.

New Nanofiltration/Reverse Osmosis Membranes

RO and NF membranes are typically cross-linked polyamide barrier layers synthesized from m-phenylenediamine (MPD) and trimesoyl chloride (TMC). Thin-film composite (TFC) membranes offer more flux at a given applied pressure. These membranes are less susceptible to microorganisms and are more stable over a wider range of pH values than cellulose acetate (CA) membranes. However, their chlorine tolerance constrains their life cycle and performance.

MPD/TMC membranes are susceptible to oxidation from chemicals such as chlorine, an inexpensive biocide widely used in water treatment. Alternative monomers that are less sensitive to chorine oxidation have been used in recent applications. As a result, disulfonated polysulfone-based layers have been reported to have better tolerance to chlorine than polyamide-based polymer layers. Increase in chlorine resistance will increase the productivity of membrane systems by directly impacting the longevity of the membrane. Increased tolerance to chlorine can also be beneficial considering the potential volume reduction of the concentrate.

Mixed matrix membranes (MMM) incorporate a combination of organic and inorganic materials. The incorporation of inorganic materials into organic RO TFC membranes started in the early 2000s and has been developing since. The use of MMM allows for the combination of benefits offered by each material. For example, the high packing density, good permeability and selectivity, and long operational record of polymeric membranes are complemented by the superior chemical, biological, and thermal stability of inorganic membranes.

Recent advancements in nanocomposite materials have given light to a new generation of TFC films integrated with functional nanoparticles. The application of super-hydrophilic and antimicrobial nanoparticles has been shown to reduce the amount of energy required for clean water production as well as the amount of effluent concentrate volume. The reductions in energy are attributed to the purification of thin-film nanocomposite (TFN) membranes as the nanoparticles actively and passively resist fouling. Titanium oxide (TiO_2) is widely used for disinfection and decomposition of organic compounds, making it an interesting material to use as antifouling coating. When coupled with ultraviolet (UV) excitation, membranes coated with TiO_2 display superior antibiofouling properties. Preliminary bench studies have shown that the application of TiO_2 onto RO membranes can reduce the volume of effluent concentrate by decreasing the amount of antifouling chemical required (Jeong et al. 2007). Furthermore, unlike other systems modifications proposed, large-scale production of TFC membranes with nanocomposite materials can be achieved without significant alterations to current manufacturing infrastructure. Application of such membranes would reduce costs by decreasing the amount of energy required and also reduce the amount of concentrate volume produced (Jeong et al. 2007).

New Electrodialysis Membranes

The early ion-exchange membranes used for ED and EDR were anion and cation exchange membranes made from crushed ion-exchange resins in an inert matrix. Over the years, these membranes were replaced by newer membranes with lower electrical resistance (i.e., higher energy efficiency) and higher scaling and fouling resistance.

In the 1980s, acrylic-based anion exchange membranes were introduced. This type of membrane is substantially more resistant to organic fouling. It has a much higher chlorine tolerance, which allowed the EDR systems to be operated with up to 0.5 mg/L of free chlorine residual. It also allowed the membranes to be chemically cleaned with up to 20 mg/L of free chlorine, resulting in more effective cleaning of severely fouled biological or organic membranes.

In 2011, a caustic-stable ion-exchange membrane was released for EDR that allows for more aggressive cleaning-in-place (CIP) of EDR stacks. This membrane, named AR908, allows the EDR stacks to be cleaned under high pH up to 13. The ability to more aggressively clean the EDR stacks is significant for wastewater treatment and other applications with high organic and biological fouling. It could reduce the need for opening the stacks and manually cleaning the membranes. It also could reduce the pretreatment requirements for EDR, so that raw water treated by cartridge filters, granular media filters, or simply secondary clarifiers could be a suitable EDR feed.

Other improvements provided by AR908 include the reduction in membrane spalling, which may result in longer membrane life. This membrane could also be used interchangeably with AR204 to allow a partial replacement of the membrane stack.

Driven by the need for selective removal of sodium (for irrigation), chloride (for effluent discharge, etc.), or nitrate, monovalent selective ion-exchange membranes are also being tested for enhanced removal of target ions from saline reclaimed water and brackish water concentrate. Conventional RO membranes cannot remove monovalent ions selectively. Generally, NF membranes have selectivity. These membranes are often looser and reject more large ions (e.g., calcium, sulfate) than smaller ions (e.g., sodium, chloride), worsening the situation. The monovalent selective IX membranes for ED/EDR (e.g., sodium selective membrane CR671 and nitrate/chloride selective membrane AR112B) potentially could remove more monovalent ions than divalent ions at lowered electrical current, which offers economical solutions to many inland salinity challenges associated with sodium adsorption ratio and whole effluent toxicity failure caused by chloride.

Other emerging membranes. Carbon nanotubes (CNTs) have fluid transport characteristics similar to water transport channels in biological membranes. High selectivity of particles allowed to pass through the membrane decreases the amount of water rejected into the concentrate stream. CNTs facilitate conveyance for water molecules while rejecting salts and pollutants (Das et al. 2014). CNTs can serve as pores in the membrane for water desalination and decontamination (Elimelech and Phillip 2011). Furthermore, CNT membranes are capable of selectively selecting and rejecting ions (functional tip-end). Therefore, CNT membranes can be used as highly trained "gatekeepers" for size-controlled separation of multiple pollutants (Das et al. 2014). Current research has been focused on further developing the fabrication and functionalization of CNT membranes. Ratto et al. (2011) patented a CNT membrane that is capable of achieving a TDS removal greater than 99 percent. Because of limited pilot testing, economical tendencies and cost analysis cannot be completed at this point.

The water transport properties of biological membranes, such as cell walls, have motivated researchers to study membranes incorporating water-selective proteins. The inclusion of water-selective properties would serve as water-directing channels. Research has shown that incorporating bacterial Aquaporin Z (AqpZ) proteins increased water transport efficiency relative to conventional RO membranes (Tang et al. 2013); the authors postulated that by using AqpZ-based biometric membranes, a permeability of two orders greater than that seen in conventional seawater RO (SWRO) systems could potentially be achieved.

Theoretically, biomimetic membranes should offer the highest permeability; however, because of the lack of extensive research, this conclusion is based solely on transport velocity measurements of water molecules through individual AqpZ channels rather than conventional membrane tests. At present, two companies are focused on developing biomimetic membranes. Aquaporin A/S is developing a two-dimensional membrane with embedded aquaporins to support pressures up to 10 bar, with flux >100 lmh. Danfoss AquaZ is developing three-dimensional membranes within which vesicles are cross-linked tightly together with space between them filled up, and the water then passes through several aquaporins; the objective of such a process is to achieve a flow that is five to ten times higher than that of conventional TFC membranes (Veerapaneni et al. 2011).

INNOVATIONS IN DESIGN AND OPERATION

Innovations in Membrane Scaling Monitoring

Mineral scaling occurs when the concentration of sparingly dissolved mineral salts such as gypsum, barium sulfate, calcium carbonate, or silica near the membrane surfaces rises above their solubility limits. As permeate recovery is increased to offset cost associated with management of concentrate, there is also an increase in the fouling propensity of brackish water desalination because of this phenomenon. As a result, sparingly soluble salts can precipitate as bulk on the membrane, leading to a significant reduction in membrane performance and membrane life. If development of scaling is detected at an early stage, membrane cleaning can be accomplished effectively through processes such as chemical cleaning (USBR 2009) or feed flow reversal (Bartman et al. 2008). These processes require advanced real-time membrane monitoring to detect the onset of mineral scaling (Bartman et al. 2011) in a timely manner.

In-situ monitoring techniques for the study of mineral scaling and use in mineral scale and fouling detection in RO systems have been based on indirect ultrasonic crystal detection (Zhang et al. 2006). Although these methods are capable of detecting the buildup of a foulant layer, they are incapable of differentiating the type of foulant (e.g., particulate deposition, bacteria, or surface-formed crystals). Furthermore, in general, there is insufficient reliability in the early-onset detection of membrane surface crystallization (Uchymiak et al. 2009, 2007).

To overcome the limitations of RO systems controller and indirect ultrasonic crystal detection, a novel ex-situ membrane monitor (MeMo) has been developed (Bartman et al. 2011). The MeMo detector is capable of delivering real-time images of a representative membrane surface, where the operating conditions are matched to the conditions in the spiral-wound element of the RO plan (Bartman et al. 2011). Through the use of an image analysis program (operating either online or in a postprocessing mode), plant monitoring systems would be capable of determining the amount of membrane surface coverage by mineral salt scaling and the number of crystals present in the observation area (representative of a spiral-wound element of the RO plan). With this monitoring approach, when mineral scale coverage reaches a predetermined threshold, a control signal can be sent to the plant control system to initiate cleaning strategies, adjustment of antiscalant dose, or adjustment of product water recovery.

Reversible RO Configuration

Conventional RO typically uses a series of membrane elements in a pressure vessel. Typically, the front-end membrane elements produce more water and are more prone to biofouling and organic fouling. In contrast, the back-end elements are more prone to scaling, as the concentration of the sparingly soluble ions increases as the feedwater flows through the concentrate channel. The reversible feature uses an advanced control system and motorized valves to allow for the feed flow to each RO membrane stage to be reversed periodically, thus making the lead element the lag element, and so on. It was demonstrated that when the reversible cycle time was every eight to twelve hours, noticeable improvements were observed, to the point that normalized net driving pressure flattened in both membrane stages, indicating a reduction in fouling. Membrane autopsy results suggest that biofouling may be reduced by applying the reversible operation feature.

Closed-Circuit Desalination Process

In traditional RO or NF processes, the amount of permeate that can be recovered by a stage of membrane elements must be limited to even the flux distribution and provide sufficient crossflow in the final stages of RO, where scaling risks are the highest in the system. In the closed-circuit desalination (CCD) RO/NF process, a high-pressure pump feeds a closed-loop comprising a single stage of membrane elements and a circulation pump. Permeate is produced at a rate equal to the flow rate of the high-pressure pump. Brine is recirculated without depressurization. When a desired recovery percentage is reached, brine is throttled out of the system, displaced by feedwater from the high-pressure pump in a once-through plug-flow sweep. Brine rejection occurs without stopping the high-pressure pump or the production of permeate. The process then returns to closed-loop operation, during which there is no brine reject stream. The initial pressure requirement of each sequence is just above the osmotic pressure of the feedwater, and the final pressure requirement is just above the osmotic pressure of the final brine—the same pressure as the running pressure of a stage of membrane elements in a traditional continuous system. The resulting average membrane feed pressure in the CCD process is lower than the feed pressure of typical RO or NF systems. CCD systems have 15–35 percent lower energy use than conventional RO systems with pressure exchangers. Their energy efficiency is higher for lower-salinity source waters (wastewater and brackish water) where the system recovery could be increased to 85–90 percent. The CCD systems, however, have lower energy efficiency for desalination of seawater or volume reduction of brackish water concentrate. Therefore, practically all full-scale applications of the CCD systems to date have been for water reclamation and low-salinity brackish water applications.

Advancement in RO Membrane Cleaning

Recent innovations for membrane cleaning include use of small air bubble cleaning suspension, use of high ionic strength cleaners, and early predictive cleaning.

Small air bubbles produce agitation and scrubbing of deposits on the membrane surface, which enhance cleaning effectiveness. High ionic strength cleaners (e.g., concentrated NaCl solution) create a small flow of permeate across the membrane to the feed side during the soaking period of cleaning, which helps dislodge foulants on the membrane surface. Applying membrane cleaning early in the fouling process before deposits can become compressed may also provide operational savings on life cycle as compared to waiting until membranes are so fouled that their productivity is reduced by 10–15 percent, as is traditionally recommended. Early cleaning makes deposits significantly less difficult to remove (Chesters et al. 2013). Most of these simple techniques can be easily applied to existing RO plants.

CIP with chemicals is most widely used to remove foulants and maintain membrane performance. However, CIP needs a downtime of frequent RO operation stoppage, resulting in reduced production and creating environmental issues related to the waste chemical disposal. The direct osmosis–high salinity (DO–HS) backwash cleaning method offers a novel approach for the RO membrane operation, which can be used to reduce the frequency or extent of CIPs required. The DO–HS process cleans the membrane without interrupting the operational process. The force required for the DO–HS cleaning process can be provided by 25 percent concentrated, activated NaCl solution, with 194 bar osmotic pressure (Liberman and Liberman 2005).

INNOVATIVE PROCESSES

Besides the mainstream treatment technologies discussed in Chapter 3, such as NF/RO, ED/EDR, chemical softening, and thermal process, several innovative desalination and concentrate treatment processes could be considered. Some of these technologies are still in the development stage; others have full-scale installations for certain industrial applications. When used for the right application, each of these technologies could offer critical benefits in achieving the overall treatment goals.

Vortex-based Anti-fouling Membrane System

Vibratory Shear Enhanced Process (VSEP®) is briefly discussed in Chapter 5. A similar innovation, the FMX technology, is briefly introduced herein. Unlike VSEP, the FMX filtration system prevents the fouling of its flat-sheet membrane modules through the use of vortex generators. The vortex-generating blades are constructed of a durable engineered plastic proven to be highly resistant to chemicals, corrosion, and temperature. These vortex generators rotate between the stationary membrane modules, creating a sustained turbulence. This turbulence prevents the settling of suspended solids upon the membrane surface, and the solids are removed from the system with the concentrate stream. This process reduces maintenance and downtime, while maintaining a steady flux relative to influent quality over long-term operation (Eum et al. 2012). The unique design of the vortex-based antifouling membrane system allows it to overcome the limitations that have long prevented the application of membrane filtration for high-solids, high-fouling industrial waste and process streams. Therefore, it is expected that FMX could be a highly promising solution for high-solid-waste, industrial wastewater—maintaining the broad removal abilities of conventional membrane systems while eliminating challenges resulting from fouling and scaling. This could also reduce the need for extensive pretreatment.

Electrodialysis Metathesis

The electrodialysis metathesis (EDM) process was patented at the University of Southern Carolina (Patent PCT/US2005/032419). This system is composed of a traditional RO system followed by the EDM process. The EDM process is discussed in detail in Chapter 5.

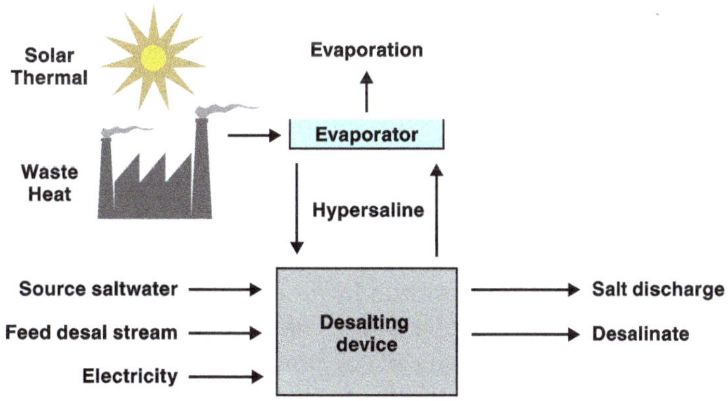

Figure 9-1 General schematic of the thermo-ionic process

Thermo-ionic Technology

Thermo-ionic technology was developed by Saltworks Technologies Inc., Richmond, British Columbia, Canada. Using a hybrid design, the technology developer incorporated thermo-ionic technology into an RO system to provide potentially low-cost concentrate management. Thermo-ionic desalination uses ionic gradient to treat the concentrate stream. The process is driven by the salt concentration gradient between hyper-saline solution and feedwater. Figure 9-1 shows a simple schematic of the process. The hyper-saline solution is produced in a special evaporative unit that operates at a temperature 10°C warmer than the ambient wet bulb temperature. Because the unit relies on salinity gradients for internal voltage generation, the net salt flux, or current density, is lower than that of EDR, requiring more membrane area (Xu et al. 2013).

Further development of thermo-ionic technology has recently been conducted by SPX Cooling Technologies, Centennial, Colo. SPX Cooling Technologies has developed an air-cooled tower potentially capable of achieving ZLD while also harvesting precipitated salts. Although this technology has great potential for application in RO concentrate treatment, it is still in the development stage; thus, further bench-scale and pilot testing must be completed before any significant recommendations can be made.

AquaSel™

AquaSel™ is a nonthermal brine concentrator (NTBC) for high water recovery. The combination of a desalination unit and a concentration unit can potentially remove barriers to water recovery arising from inorganic scale fouling. The concentration unit is then specifically designed and operated to reduce the volume of the waste stream from the system. The heart of the process is an EDR system that can be adapted to further concentrate the RO reject stream by targeting specific concentrate applications.

After it has been desalted, the EDR product water is re-introduced into the RO system and the reject concentrate is discharged into an engineered precipitation tank, where the precipitated solids are allowed to settle. As a result, dry salt cakes are produced in the precipitation tank. With a simple retrofit, any RO process would be augmented to only discharge less than 5 percent of the feedwater (Khos et al. 2016).

Adsorption Desalination

Adsorption desalination (AD) is an emerging, environment-friendly technology for potable water production using a highly porous silica gel that can be regenerated by low-temperature waste heat or renewable solar heat. It is a thermal-based evaporative process in which vapor is removed from saline or brackish water via evaporation in a single component environment, caused by the high affinity of vapor uptake onto the pores of the adsorbent material. Further technical information was not available at the time this manual was developed.

Customized hybrid systems. Concentrate treatment can be relatively complex, and no single "silver bullet" technology is available to address all issues that typically need to be addressed. Hybrid systems, customized based on the site-specific water quality and treatment goals, might result in lower energy consumption, higher recovery, mineral recovery, and optimized performance. This involves maximizing chemical conditioning, leveraging organic and inorganic pretreatment technologies to control scaling and fouling, and maximizing the recovery of the secondary desalination process.

As discussed in Chapter 5, one hybrid example is the High Efficiency Reverse Osmosis (HERO™) process. It consists of optional chemical conditioning using alkali and acid, softening using weak acid cation exchange resins, degasification for CO_2 removal, and RO to achieve high water recovery, typically in the range of 90–98 percent. See Chapter 5 for more details.

EDR and slurry precipitation and recycling RO (SPARRO) technologies were recently tested through a USBR Desalination and Water Purification Research Program (DWPR) project. Approximately 8–10 gpm of concentrate from an existing groundwater RO facility (the Temescal Desalter, owned and operated by the City of Corona, Calif.) was used for the four-month testing. In the combined process, the SPARRO unit is connected into the EDR concentrate loop to reduce the scaling potential in the concentrate loop. This improves EDR performance and allows the recovery of the EDR system to be increased. The SPARRO technology allows for direct desalting of feedwater that has high concentrations of calcium sulfate, without the need for an intermediate softening step. A byproduct of the SPARRO process is high-quality gypsum solids, which may be sold as a useful product.

Another example is the recently completed concentrate management study co-funded by the Sub-regional Operation Group (SROG, including Glendale, Mesa, Phoenix, Scottsdale, and Tempe, Ariz.), the WateReuse Research Foundation, and the USBR. This study demonstrated the feasibility of reducing tertiary reclaimed water RO concentrate volumes and improving overall RO recovery from 85 to 94 percent by using EDR with organic matter pretreatment (ozone and biologically activated filtration [BAF]), and to 98 percent with the addition of inorganic pretreatment as well (e.g., ion exchange or lime softening). EDR reliably recovered 62 percent of the RO concentrate with organic pretreatment alone, and 89 percent with both organic and inorganic pretreatment. This means that the proposed concentrate treatment schemes could improve the overall recovery from 85 to 94 and 98 percent, respectively. This demonstration project delivered new brine management solutions that have reshaped inland desalination and concentrate management. The overall recovery and the total water costs associated with the concentrate management system (i.e., ozone, BAF, IX plus EDR with regular membranes) provided ~40 percent cost savings in the total water costs compared to the baseline alternative using brine concentrators, with only a 0.7 percent difference in overall recovery. Compared to secondary RO with organic pretreatment and lime softening, the demonstrated solution offers ~30 percent cost reduction while achieving a slightly higher overall recovery.

ENERGY EFFICIENCY ENHANCEMENT

Several strategies can be considered for enhancing energy efficiency associated with inland desalination and concentrate management. Among the frequently applied strategies are use of energy recovery devices (ERDs), collocation of the desalination plant with the power plant, and development of renewable energy.

Energy Recovery

Because of low feedwater TDS and relatively low concentrate flow, the applications of ERDs in brackish water RO desalination may be ambiguous without some analysis. To implement ERDs successfully, a careful analysis of the return on investment in the ERD (based on life-cycle cost analysis) must be completed. Because of variability in concentrate stream, the ERD must be designed for a wide range of flow rates. Treatment of concentrate with higher TDS offers more opportunities for energy recovery than does brackish water desalination. For the purpose of brackish water RO desalination, two types of ERD are typically considered: centrifugal turbocharger and isobaric ERD (Drak and Adato 2014).

The turbocharger is a two-part hydraulic device that consists of a pump section and a turbine section, both containing a single-stage impeller or rotor. As liquid concentrate flows into the turbine section of the turbocharger, the turbine rotor extracts the hydraulic energy from the concentrate and converts it into mechanical energy. The pump then introduces additional hydraulic energy to the feedstream from the mechanical energy produced (Drak and Adato 2014). The turbocharger is entirely powered by the reject stream. The overall efficiency of the turbocharger, which ranges from 50 to 80 percent, is a function of the efficiency of the turbine and the pump impeller (Manth et al. 2003).

One option to eliminate the efficiency losses that arise from multiple energy conversion steps (hydraulic to mechanical and mechanical to hydraulic) is to use the isobaric ERD. Using the principle of positive displacement, energy transfer from a high-pressure concentrate stream to a low-pressure feedstream can be achieved. As the feed and concentrate streams are in direct contact, there is an increase in salinity in the feedstream to membranes. The increase in salinity creates an increase in the operational pressure of approximately one to two bars (Stover 2007, 2004). The piston-type ERD operates similarly to the rotary type; however, instead of a rotor, positive displacement pistons are used. Two separate pressure vessels are used to create the interaction between high-pressure concentrate and low-pressure feedwater. In one vessel, the high-pressure concentrate displaces low-pressure feedwater, while in the second vessel, the low-pressure feedwater displaces the high-pressure concentrate (Schneider 2005). A typical setup of a two-stage RO system with an isobaric ERD can be seen in Drak and Adato (2014).

Hybridization of Desalination and Power Plants

The water and energy nexus demands more collaboration among water utilities and power plants. A hybridized desalination plant incorporates desalination systems used in combination with an inland power plant. There are many advantages to this combination. A main advantage is the possibility of maintaining a constant feedwater temperature using the available waste heat from thermal concentrate discharge as well as effluent heat discharge from the power plant. More details on benefits and challenges of collocation of desalination and power plants are provided elsewhere (Voutchkov 2013).

Use of Renewable Energy

It is crucial that the use of renewable energy for brackish water desalination be implemented for two reasons: (1) the price of fossil fuels has considerably increased over the past decade, a trend that may continue for years to come; and (2) to reduce the carbon footprint. The use of renewable energy in brackish water desalination has the potential to be implemented through the use of hybrid systems and renewable sources of energy such as geothermal, wind, and solar. With improvements in membrane technology and increasing fossil fuel prices, the application of salinity-gradient energy production using pressure-retarded osmosis or reverse electrodialysis technologies may also become more attractive in the future. A comprehensive report on the use of renewable energy for desalination is provided in a recently published World Bank report (2012).

IMPROVED SALT RECOVERY AND BENEFICIAL USES OF CONCENTRATE

Significance and Challenges of Salt and Nutrient Recovery

Recently, new technologies have emerged that recover the salt content of concentrate. These include ZLD, chemical and biological precipitation processes, pelletized softening, the SAL-PROC™ process, selective salt recovery, and advanced solar dryer.

Pellets produced as a result of chemical precipitation processes could be used as construction materials and animal feed supplements and for post-treatment. Furthermore, with the use of some innovative biological processes, it is possible to precipitate common metals out of effluent concentrate stream such as calcium, magnesium, and potassium. Precipitation of such metals would also reduce the need for invasive mining techniques, thus further reducing the environmental footprint and water consumption of mining industries. Moreover, selective precipitation (removal) of the carbonate series may help reduce the CO_2 footprint of desalination.

The importance of the development of salt recovery systems is beyond providing a cost-effective method for concentrate disposal. The separation and marketing of salts is a direct step toward developing an environmentally sustainable solution where water recovery is maximized and landfill use is minimized.

Zero Liquid Discharge Systems

ZLD processes have been discussed in previous chapters. Several ZLD technologies have been successfully implemented in industrial water treatment processes. However, the ZLD concept is relatively new in the application to large-scale RO systems. Developmental technologies such as WAIV (wind aided intensified evaporation), Dewvaporation, and Salt Solidification and Sequestration are being tested to provide ZLD in large-scale RO systems.

Pelletized Lime Softening Process

To maximize water recovery, technologies such as pelletized lime softening (PLS) are being evaluated to improve the recovery rate of such ZLD systems to about 92–95 percent. Studies have shown that it is feasible to achieve a 94 percent overall recovery on brackish surface water and groundwater with the implementation of a conventional softening intermediate concentrate chemical stabilization (ICCS) reactor. ICCS technologies are implemented in between the primary and secondary RO systems to precipitate sparingly soluble ions such as silica, calcium, and barium. PLS ICCS works through the addition of microsand in an up-flow reactor. The sparingly soluble salts previously mentioned in this section will react with the microsand and precipitate as "pellets" (Shi and Liu 2005).

The PLS system provides a reduction in the chemical footprint and the volume of residuals produced as a result of ICCS. Waterborne emissions of nitrate and ammonium from pellet systems were shown to be approximately 46 and 47 percent, respectively, less than that from a conventional softening system (Luo and Wan 2011). Furthermore, pilot studies have shown virtually complete softening with no conceivable post-softening problems.

Table 9-1 Water quality before and after pelletized treatment

	Well 6		Well 3	
Parameter	Raw	Finished	Raw	Finished
pH	7.1–7.8	8.2–8.9	7.3	8.0–8.4
Hardness (mg/L)	188–210	60–72	212–253	53–81
Iron (mg/L)	0.51–0.96	0.03–0.10	0.07	0.00
Manganese (mg/L)	0.28–0.36	0.015–0.019	0.28	0.018

Source: Snedecor et al. 2008

A recent study established that for groundwater conditions similar to those found in the Phoenix, Ariz., area, PLS ICCS lowers environmental impact because of a reduction in the consumption of chemicals, energy, and residual handling requirements (He et al. 2011).

Pilot studies have shown that PLS can reduce the hardness range and the iron and manganese concentrations. In a case study, the quantities in Table 9-1 were reported.

When compared to conventional lime ICCS, the pelletized lime ICCS provided a reduction in four impact categories. Two separate pilot studies were conducted to provide a comparable report on the proposed ICCS technologies (Wan et al. 2011). The conventional softening ICCS pilot study was conducted in the City of Phoenix, Ariz., at the Western Canal WTP as part of both the Eastern Municipal Water District Desalination Recovery Enhancement and Concentrate Management Study and the South Florida Water Management District Water Desalination Concentrate Management and Piloting. The pelletized softening ICCS pilot study was conducted in the City of Tracy, Calif., as part of both the Western Municipal Water District Arlington Desalter Concentrate Pilot Study and the Tracy Concentrate Pilot Study.

Residual pellets can be used as nutrient sources in animal feed (chicken factories), in the cement (aggregate) and metal industries, and in road construction (asphalt). Residual pellets contain some or all of the key elements that cement plants add during the cement manufacturing process. The addition of these residuals could decrease the total volume of supplemental materials that cement manufacturers would have to purchase, decreasing invasive mining techniques that are currently in practice. The "dry" nature of the pellets makes them attractive to cement manufacturers as this reduces the need for any dewatering processes.

SAL-PROC™ Process

Salt Solidification and Sequestration (SAL-PROC™) is a patented process (Geo-Processors USA, Inc., Glendale, Calif.) where valuable salts and chemical compounds are extracted from inorganic saline waters. The saline feedwater is initially pretreated using a solar evaporation or desalination process supplemented by conventional mineral and chemical processing steps. The steps could lead to a significant reduction in the salt load and volume, minimizing the requirements for effluent discharge. This process can be considered when the chemical composition requires multiple stages of reaction.

Geo-Processors USA has developed a model that consists of two subsystems, including one or more selective salt recovery steps coupled with RO desalination, thermo-mechanical concentrate concentration, and crystallization steps. The model allows for selection of an appropriate ZLD process scheme. The ZLD processes use multiple steps and incorporate lime and soda ash to precipitate mixed salts. The overall system has been shown to potentially recover the entire concentrate flow and generate high-quality water as a result. However, the incorporation of desalting technologies is required to reduce the volume significantly while highly concentrating water entering the SAL-PROC.

Pilot testing has shown that a number of saline waste streams can potentially be converted into marketable products while achieving ZLD (Jibril and Ibrahim 2001). Typical recovered marketable products include but are not limited to gypsum-magnesium hydroxide, magnesium hydroxide, sodium chlorite, calcium carbonate, sodium sulfate, and calcium chloride. Through the use of SAL-PROC, Jibril and Ibrahim (2001) produced, at laboratory scale, sodium bicarbonate, sodium carbonate, and ammonium chloride from concentrated sodium chloride solutions by first ammoniating the solution and then bubbling carbon dioxide through the concentrate.

SAL-PROC is not a stand-alone concentrate stream treatment technology; it is a process recovery system. The suitability application largely depends on effluent water quality and type of application. In some cases, RO concentrate from water reuse facilities might not be compatible with this system because wastewater can contain organic, toxic, and hazardous material. Consequently, SAL-PROC systems cannot be incorporated into wastewater treatment facilities. However, the production of marketable products and ZLD would allow developing/stand-alone communities to offset the cost of RO systems. Because of proprietary software and technology, cost data for SAL-PROC are not available.

Selective Salt Recovery Techniques

In the gas mining industry, there is a process to recover sodium chloride crystals and sodium carbonate decahydrate crystals from saline water. The feed concentrate for the process is a byproduct of a gas mining operation (coal seam gas) that produces highly concentrated salt water. Selective salt recovery techniques refer to methods or processes for selectively recovering salts from a mixed salt concentrate—in particular, the recovery of sodium chlorine crystals and sodium carbonate from a concentrate that is generally rich in these salts. The application of such a system should be considered once an initial study is conducted to determine what types of recoverable salts are available in the concentrated concentrate.

An initial pre-concentration process is carried out where the concentrate is concentrated; in the process, carbon dioxide is removed from the concentrate. Some of the sodium bicarbonate is converted into sodium carbonate initially. During the process, the concentrated effluent concentrate is pumped into a sodium chloride crystallizer where the concentrate is heated and further concentrated to form sodium chloride crystals. The sodium chloride crystals produced are separated from the concentrate to yield a salable salt product. As a result of sodium chloride crystallization, the mother liquor (effluent concentrated concentrate post-sodium chloride crystallization) is then directed to a sodium carbonate decahydrate crystallizer where the liquor is cooled and concentrated, resulting in the formation of sodium carbonate decahydrate crystals and a second mother liquor. The second mother liquor is split into two streams: one stream is directed back to the sodium chloride crystallizer for a second time, while the other stream is wasted or further treated using other technologies.

The recovery of the salts may require beneficiation to maximize commercial value. For example, it may be appropriate to improve the quality of sodium chloride by dissolution and re-crystallization—a process often done with sea salt. Applying processes such as the one shown in Figure 9-2 may broaden the applications of the sodium carbonate decahydrate crystals. Such processes would form dense soda ash or sodium bicarbonate. Soda ash has been shown to have applications as sodium silicates in a cement sealer. Furthermore, sodium carbonate is applicable to various industries such as food production and pool maintenance.

The previously described processes to recover sodium chloride and sodium carbonate decahydrate from mixed concentrate have various advantages. Compared to known processes, this process generates little to no waste material. Furthermore, compared to

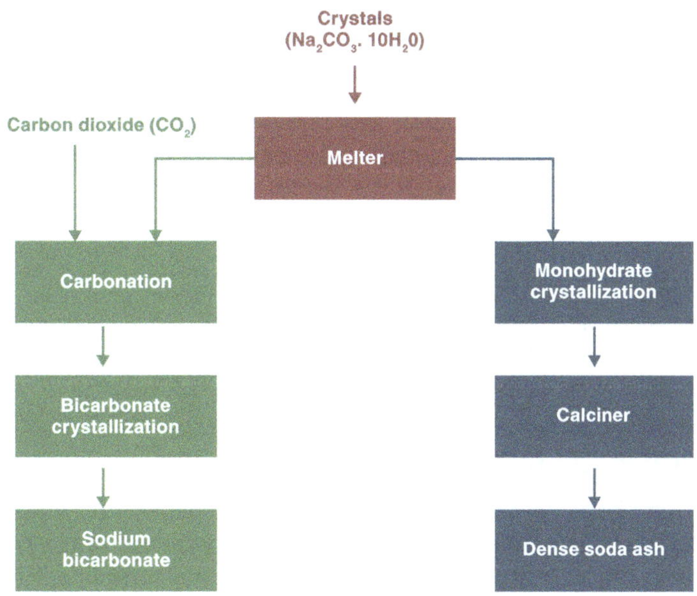

Adapted from Rittof and Rieke 2013

Figure 9-2 Selective salt recovery from sodium carbonate crystals

known processes, this process requires little or no reagents to recover the salts. Finally, the process does not require evaporation ponds; this is crucial in the application as minimal space is required for salt recovery.

Advanced Solar Dryer Process

The main objective of the advanced solar dryer (ASD) process is to focus on the exploitation of sodium chloride as a byproduct resulting from the effluent treatment process through concentrate concentration in a solar passive dryer. The main objective is to reduce the cost of clean water production by decreasing production costs and increasing the revenue from sodium chloride production. An ASD is a prototype system developed as a result of the study of a numerical model describing dryer operation under given yearly climate conditions. This particular methodology allowed researchers to gain a deeper knowledge of the design and identification of further evaporation enhancement technologies. The prototype under study (Collares-Pereira et al. 2003) was developed with a 20 m^2 evaporation area for testing under real concentrate evaporation.

Other Chemical Processes

The chemical components in the concentrate stream can be used for additional applications after being extracted through physical or chemical processes. Mekorot Water Company operates a dual-purpose plant in Israel that is able to produce desalinated water and high-quality table salt (Ravizky and Nadav 2007). The desalination plant uses a mixed feed of 80 percent seawater and 20 percent brackish water RO (BWRO) concentrate from adjacent plants. Both feeds are blended together and then placed in multiple evaporation ponds before being processed at the salt factory of the company. The production of salt has not been limited to RO concentrate. Tanaka et al. (2003) reported that using a hybrid system composed of ED to treat concentrate from an SWRO plant as feed for salt production may provide significant energy reduction, up to 20 percent, and be more advantageous than using seawater. The preferential selectivity of monovalent ions over multivalent ions (e.g., Na^+ over Ca^{2+}, Cl^- over SO_4^{2-}) required the use of special grade monovalent permselective IX membranes in ED stacks. As a result, Tanaka was able to produce food-grade sodium chloride from the concentrate stream.

A bench study conducted by Badruzzaman et al. (2009) showed that through the application of ED in combination with bipolar membranes (BMED), it is possible to recover acid, base, and hypochlorite, which could be used directly onsite. An economic evaluation showed that this approach might be feasible for inland wastewater reuse facilities, which have limited access/options for concentrate disposal. In this process, monopolar cation- and anion-exchange membranes are installed together with bipolar membranes in alternating series in an ED stack (Xu et al. 2013).

For systems where multiple salts are recovered and where there is a wide range of solubility, the processing may involve multiple stages of concentration and treatment steps that can recover salts from lowest to highest solubility.

Biological Precipitation Process

Researchers have been able to isolate bacterial species that are capable of accumulating minerals from their environment over a certain period of time. As the bacteria metabolize, they create an electrical charge, thus attracting metals from their local environment. These metals accumulate as material on the surface of the bacteria. Biomass can be accumulated and minerals can be extracted.

Recoverable Salts

An analysis conducted by Mickley in 2009 showed that there are many applications for the major recoverable salts (Table 9-2), and many of the salts have sufficient value to make their recovery economically attractive. The recoverable products are largely site-specific, thus their recovery and marketability depends largely on the following factors: volume of concentrate produced, quality of water (type of dissolved salts), quality (form and purity) of salts recovered, reliability and consistency of salt quality, local applications for recovered salts (potential markets), and existence and reliability of the local market and combined income from sale of the different salts. Each site will require in-depth analysis to determine these and other issues before development of concept. Areas such as the southwestern United States would benefit from these products as systems capable of recovering salts would provide alternative sources of revenue, making inland desalination more economically attractive.

Sample Case Studies for Beneficial Uses of Pellets for Soil Applications

Case study 1: Soil remediation. Pellets produced as a byproduct of ZLD processes may be used in soil remediation. An example of soil remediation is through the use of calcium chloride for sodic (high sodium adsorption ratio) soils instead of gypsum, which is cheaper and well accepted. Rainfall on sodic soils cannot soak into the soil and is therefore lost by evaporation. Because of these characteristics, the soil does not support agriculture, and rainfall is lost to the local areas as the water cannot enter lower layers. By adding calcium salts, it is possible to change the hydraulic conductivity of the soils to produce increased rates of infiltration and reduce the amount of rainfall lost, in turn significantly increasing the supply of usable water (Mickley 2008).

Current practices for sodic soil remediation include application of gypsum salt. More effective soil remediation techniques can be achieved through the use of calcium and magnesium chloride for sodic soils (Mickley 2008). These products, however, are more costly than gypsum. Even though more expensive chemicals may be more efficient and perform at a higher level, farmers remain reluctant to pay more money for remediation.

Table 9-2 List of recoverable salts and their potential application

Chemical Formula	Name	Application Areas
$CaCO_3$	Calcium carbonate	Paper coating pigment, fillers for plastics and rubbers, special inks, paints, and sealants
$CaSO_4 + 2H_2O$	Gypsum	Remediation of sodic soils, manufacture of building products
$CaSO_4 + 2H_2O + Mg(OH)_2$ slurry	Gypsum magnesium hydroxide	Wastewater treatment, pH buffering, soil conditioner for sodic soils
$CaCl_2$ (liquor)	Calcium chloride	Dust suppression, road base stabilization, sodic soil remediation, cement and concrete stabilizer, construction industry
$KNaSO_4$	Glacerite	Potassium fertilizer
$Mg(OH)_2$ slurry	Magnesium hydroxide	Water and wastewater treatment, environmental, animal stock feed, feedstock for magnesium metal production, fire retardant and refractories, acid neutralizer
$4MgCO_3 + Mg(OH)_2 + 4H_2O$	Magnesium carbonate light	Fire retardant, feedstock for magnesium production, filler for paper manufacturing, rubber, and paint
NaOH	Caustic soda	Many applications industrially, including basic feedstock for chemical processes; pH adjustment
NaCl	Halite	Food and industrial processes, chloralkali production, bulk salt supply
Na_2CO_3	Soda ash	Water treatment, chemical industry
Na_2SO_4	Thenardite	Surfactants manufacturing, detergent manufacturing, glass manufacturing, remediation of calcareous soil
NaOCl	Sodium hypochlorite	Disinfection, chemical industries, pool chlorine, water pretreatment
$NaClO_3$	Sodium chlorate	Paper bleaching, chemical industries
NH_4Cl	Ammonium chloride	Fertilizers, metal work, expectorant in medicines, yeast nutrient in bread making
$NaHCO_3$	Sodium bicarbonate	Fire retardant/extinguisher, food processes as baking soda, neutralization of acids and bases

Adapted from www.geo-processors.com

Case study 2: Dust suppression. Calcium chloride and magnesium chloride are two types of salts that are currently being used for dust suppression. Moisture keeps fine particles in unpaved roads and construction projects together by coating and binding them. Calcium chloride is able to absorb large amounts of water and, because of high surface tension, retain it, reducing evaporation. These characteristics enable calcium chloride to coat soil particles with a strong, thin film of moisture that reduces friction between particles, increasing compaction rate. Once the particles are compacted, the surface tension creates a cohesive force that is able to hold the consolidated base together and performs a soil stabilization function.

The market for calcium chloride and magnesium chloride exists for this application, but the product will need to be tested to ensure that it meets established product specifications. The ideal application for this use of concentrate would be implemented for the life of the desalination plant.

Case study 3: Hot-mix asphalt (HMA) production. Traditionally, lime fines have been used to treat HMA ingredients to improve the long-term interactions (bonds) between the asphalt cement and aggregates. An example of such an improvement is the reduced deformation or "softening" of the asphalt over time, which is achieved by combining lime fines with either the aggregate or asphalt cement before preparation of the HMA mix. Though adding lime fines into HMA using current methodologies has been beneficial, problems such as clumping of the lime fines and the fines becoming airborne have occurred. The fines becoming airborne makes it difficult to accurately measure or control the amount of lime being used in the mix and also contaminates the environment.

It would be advantageous to produce a lime product in a form other than fine powder due to the chronic health effects associated with its ingestion. Producing lime pellets through pelletized softening would be beneficial as it would provide a method of processing lime into a more manageable form so as to improve their use in asphalt manufacturing, paving, and in soil conditioning with minimal health and cost concerns associated with small particulates. Lime pellets produced as a result of pelletized lime softening encase microsand particulates, providing an additional source of "fine" aggregate and therefore reducing the need for virgin material during HMA production.

REFERENCES

Badruzzaman, M.; Oppenheimer, J.; Adham, S.; and Kumar, M. 2009. Innovative Beneficial Reuse of Reverse Osmosis Concentrate Using Bipolar Membrane Electrodialysis and Electrochlorination Processes. *Journal of Membrane Science,* 326(2):392–399.

Bartman, A.R.; Lyster, E.; Rallo, R.; Christo, P.D.; and Cohen, Y. 2011. Mineral Scale Monitoring for Reverse Osmosis Desalination via Real-Time Membrane Surface Image Analysis. *Desalination,* 273(1):64–71.

Bartman, A.R.; McFall, C.W.; Christofides, P.D.; and Cohen, Y. 2008. Model-Predictive Control of Feed Flow Reversal in a Reverse Osmosis Desalination Process. *Journal of Process Control,* 19(3):433–442.

Chesters, S.P.; Armstrong, M.W.; Fazel, M.; Wilson, R.; and Golding, D.A. 2013. RO Membrane Cleaning, Past, Present, Future—Innovations for Improving RO Plant Operating Efficiency. Proc. The International Desalination Association World Congress on Desalination and Water Reuse 2013, Tianjin, China.

Collares-Pereira, M.; Mendes, J.F.; and Horta, P. 2003. Advanced Solar Dryer for Salt Recovery From Brine Effluent of Desalination MED Plant. Proc. ISES Solar World Congress 2003, Göteborg, Sweden.

Das, R.; Ali, M.E.; Hamid, S.B.A.; Ramakrishna, S.; and Chowdhury, Z.Z. 2014. Carbon Nanotube Membranes for Water Purification: A Bright Future in Water Desalination. *Desalination,* 336:97–109.

Drak, A. and Adato, M. 2014. Energy Recovery Consideration in Brackish Water Desalination. *Desalination,* 339(1):34–39. doi:10.1016/j.desal.2014.02.008

Elimelech, M. and Phillip, W.A. 2011. The Future of Seawater Desalination: Energy, Technology, and the Environment. *Science,* 333(6043):712–717

Eum, Y.J.; Min, J.H.; Chan, A.; Limke, J.C.; Rhu, D.H.; Kim, J.K.; and Hwang, H.J. 2012. Recovery and Treatment of Anaerobic Digestion Effluent and Hog Manure with High Solid and High Density Using Vortex Generating Membrane System. Proc. Water Environment Federation WEFTEC 2012.

He, C.; Gross, M.; Westerhoff, P.K.; Fox, P.; and Li, K. 2011. Comparing Conventional and Pelletized Lime Softening Concentrate Chemical Stabilization. Denver: Water Research Foundation.

Jeong, B.H.; Hoek, E.M.V.; Yan, Y.; Subramani, A.; Huang, X.; Hurwitz, G.; Ghosh, A. K.; and Jawor, A. 2007. Interfacial Polymerization of Thin Film Nanocomposites: A New Concept for Reverse Osmosis Membranes. *Journal of Membrane Science*, 294(1–2):1–7.

Jibril, B.E. and Ibrahim, A. 2001. Chemical Conversions of Salt Concentrates From Desalination Plants. *Desalination,* 139(1–3):287–295.

Khos, G.; Reeve, M.; and Juby, G. 2016. Desalination and Water Purification Research and Development Report No. 195: AquaSel Technology Pilot-Scale Demonstration Menifee Desalter. Denver: US Bureau of Reclamation.

Liberman, B. and Liberman, I. 2005. RO Membrane Cleaning—Replacing Membrane CIP by Direct Osmosis Cleaning. *Desalination & Water Reuse,* 15:28–32.

Luo, J. and Wan, Y. 2011. Effect of Highly Concentrated Salt on Retention of Organic Solutes by Nanofiltration Polymeric Membranes. *Journal of Membrane Science*, 372(1–2):145–153.

Manth, T.; Gabor, M.; and Oklejas, E. 2003. Minimizing RO Energy Consumption Under Variable Conditions of Operation. *Desalination,* 157(13):9–21.

Mickley, M. 2009. Desalination and Water Purification Research and Development Program Report No. 155: Treatment of Concentrate. Denver: US Bureau of Reclamation.

Mickley, M. 2008. *Survey of High-Recovery and Zero Liquid Discharge Technologies for Water Utilities*. http://www.waterrf.org/Pages/Projects.aspx?PID=4073 (accessed Feb. 20, 2019).

Ratto, T.V.; Holt, J.K.; and Szmodis, A.W. 2011. Membranes With Embedded Nanotubes for Selective Permeability. US Patent 7993524, filed Feb. 4, 2010, and issued Aug. 9, 2011.

Ravizky, A. and Nadav, N. 2007. Salt Production by the Evaporation of SWRO Concentrate in Eilat: A Success Story. *Desalination,* 205(1):374–379.

Rittof, T.J. and Rieke, J.F. 2013. Selective Salt Recovery From Mixed Salt Brine. US Patent 8603192, filed Aug. 23, 2012, and issued Dec. 10, 2013.

Schneider, B. 2005. Selection, Operation and Control of a Work Exchanger Energy Recovery System Based on the Singapore Project. *Desalination,* 184(1–3):197–210.

Shi, J. and Liu, C.R. 2005. On Predicting Softening Effects in Hard Turned Surfaces—Part I: Construction of Material Softening Model. *Journal of Manufacturing Science and Engineering,* 127(3):476–483.

Snedecor, J.; Peters, T.; and Larsen, M. 2008. *Pellet Softening: Hardness, Iron and Manganese Removal.* http://www.pnws-awwa.org/uploads/PDFs/conferences/2008/AWWA2008%20-%20Janet%20 Snedecor.pdf (accessed Sept. 28, 2014).

Stover, R.L. 2007. Seawater Reverse Osmosis With Isobaric Energy Recovery Devices. *Desalination,* 203(1–3):168–175.

Stover, R.L. 2004. Development of a Fourth Generation Energy Recovery Device. A 'CTO's Notebook.' *Desalination,* 165:313–321.

Tanaka, Y.; Ehara, R.; Itoi, S.; and Goto, T. 2003. Ion-Exchange Membrane Electrodialytic Salt Production Using Concentrate Discharged From a Reverse Osmosis Seawater Desalination Plant. *Journal of Membrane Science,* 222(1–2):71–86.

Tang, C.Y.; Zhao, Y.; Wang, R.; Hélix-Nielsen, C.; and Fane, A.G. 2013. Desalination by Biomimetic Aquaporin Membranes: Review of Status and Prospects. *Desalination,* 308:34–40.

Uchymiak, M.; Bartman, A.R.; Daltrophe, N.; Weissman, M.; Gilron, J.; Christofides, P.D.; Kaiser, W.J.; and Cohen, Y. 2009. Brackish Water Reverse Osmosis (BWRO) Operation in Feed Flow Reversal Mode Using an Ex Situ Scale Observation Detector (EXSOD). *Journal of Membrane Science,* 341(1):60–66.

Uchymiak, M.; Rahardianto, A.; Lyster, E.; Glater, J.; and Cohen, Y. 2007. A Novel RO Ex Situ Scale Observation Detector (EXSOD) for Mineral Scale Characterization and Early Detection. *Journal of Membrane Science,* 291(1–2):86–95.

USBR. 2009. *Optimization of Chemical Cleaning of Organic-Fouled Reverse Osmosis Membranes*. https://www.usbr.gov/research/dwpr/reportpdfs/report141.pdf (accessed Feb. 19, 2019).

Veerapaneni, V.; Bond, R.; Dachille, F.; and Hays, B. 2011. Emerging Desalination Technologies—An Overview. Proc. 26th WateReuse Symposium, Phoenix.

Voutchkov, N. 2013. *Desalination Engineering: Planning and Design*. New York: McGraw Hill.

World Bank. 2012. Renewable Energy Desalination: An Emerging Solution to Close the Water Gap in the Middle East and North Africa (English). Washington: World Bank.

Xu, P.; Cath, T.Y.; Robertson, A.P.; Reinhard, M.; Leckie, J.O.; and Drewes, J.E. 2013. Critical Review of Desalination Concentrate Management, Treatment and Beneficial Use. *Environmental Engineering Science*, 30:8.

Zhang, Z.; Bright, V.M.; and Greenberg, A.R. 2006. Use of Capacitive Microsensors and Ultrasonic Time-Domain Reflectometry for In-Situ Quantification of Concentration Polarization and Membrane Fouling in Pressure-Driven Membrane Filtration. *Sensors and Actuator B*, 117(2):323–331.

Index

Note: *f.* indicates figure; *t.* indicates table

A

acids, 17, 46, 48
acrylic-based anion exchange membranes, 189
activated carbon, 141
adsorption desalination, 193–194
advanced solar dryer (ASD), 199
air-stripping towers, 50–51
alkaline, 46
Altela system, 136
aluminum, 16–17
anion membrane, 54, 54*f.*
antiscalants, 17, 47, 62, 83
AqpZ-based biometric membranes, 190
Aquaporin Z proteins, 190
AquaSel™ system, 31, 193
Authority to Operate, 159

B

Barr Lake & Milton Reservoir system, 139
bicarbonate, 20–21, 88
biofouling, 18, 49
biological precipitation process, 200
boron, 88
Boulder Zone, 140
brackish groundwater, 2, 14–18, 24
  depth, in the United States, 3*f.*
  desalination, 35–67
  quality variations, 15*t.*
  regulatory, safety, operational, and environmental issues, 131–152
  types of desalination technologies, 25–26, 27*t.*
  water quality parameters affecting desalination, 15–16
brine concentrators, 107–110, 107*f.*, 146, 177*t.*
  crystallizer and, 110*f.*
  feed TDS and maximum concentrated brine effluent, 108*t.*
  MVC equipment costs, 128*t.*
  Ocotillo Brine Reduction Facility, 158

C

capacity, 6
carbon nanotubes, 189–190
cartridge filters, 48
  horizontal vessels used as RO pretreatment, 49*f.*
case studies
  beneficial uses of pellets for soil applications, 200–202
  Chino Desalter Authority Concentrate Reduction Facility, 182 186
  City of Chandler Ocotillo Brine Reduction Facility, 155–160
  City of Palm Coast Zero Liquid Discharge Reverse Osmosis Concentrate Management, 160–164
  facilities, 156*t.*
  Fargo Wastewater Treatment Facility, Effluent Reuse Facility, 167–171
  Irwin Water Works, Army National Training Center, Fort Irwin, Calif., 164–167, 165*f.*
  Kay Bailey Hutchison Desalination Plant, El Paso, Tex., 171–174, 171*f.*
  Laguna County Sanitation District Deep-Well Concentrate Disposal, 179–182
  Zero Liquid Discharge Water Treatment Facility, 175–178
casing, 78
cation membrane, 53–54, 53*f.*
caustic soda, 46
caustic-stable ion-exchange membrane, 189
cellulose acetate, 36–37
centrifugal devices, 43
chemical safety, 47
chemical storage and feed facility, 158
chemicals, 128, 129*t.*
Chino Desalter Authority Concentrate Reduction Facility, 182–186
  anticipated chemical usage, 185*t.*
  concentrate reduction facility overview, 182–185
  costs, 186

facility location map, 183*f.*
operations experience, 185
pellet reactors and secondary RO, 185*f.*
primary RO characteristics, 184*t.*
process schematic, 183*f.*
RO concentrate management and permit, 185
secondary RO characteristics, 184*t.*
water quality, 184*t.*
chlor-alkali process, 141
chloride, 88
chlorine, 63, 66
citric acid, 46
City of Chandler Ocotillo Brine Reduction Facility, 155–160
operation challenge, 160
process challenges, 157–158
process flow schematic, 156*f.*
process overview, 158–159, 159*f.*
process permitting requirements, 159
project delivery, 159–160
City of Palm Coast Zero Liquid Discharge Reverse Osmosis Concentrate Management, 160–164
cost, 163
facility summary, 160
membrane system overview, 160
operations experience, 162–163
RO concentrate management and permit, 161–162
Class A pan coefficient, 81
Class I wells, 79, 135
Class II wells, 79
Class V wells, 79, 135
clean-in-place (CIP) system, 45–46, 62, 157
Clean Water Act, 72, 74, 133, 134–135
closed-circuit desalination (CCD), 191
closed circuit reverse osmosis (CCRO), 104–105, 104*f.*
colloidal silica, 17
communication
implementation of solution, 149–150
materials, 150
communication plan, 150
Comprehensive Environmental Response, Compensation, and Liability Act (CERCLA), 133, 136
concentrate, 36, 78
beneficial uses of, 196–202
and raw water, 143*t.*
concentrate density, 142
concentrate disposal, 25, 140
discharge options, 71–91
deep-well injection, 76–80
discharge to sewer, 75–76
discharge to surface water, 72–74
enhanced evaporation systems, 84–87
evaporation ponds, 80–83
irrigation, 87–91
concentrate management
current issues, 131–132, 132*t.*
desalination cost, 119–130
annual costs, 127–130
construction costs, 122–127
guidelines for cost estimate classes, 121*t.*
major treatment cost by category, 120*t.*
methods for estimating Class 4 and Class 5 construction costs, 121–122
overall cost considerations, 120–121
future permitting considerations, 141
technological trends, salt recovery, and beneficial uses, 187–202
industry trends, 188
concentrate management options, 7–11, 8*t.*, 13
disposal options and number of plants per state, 9*t.*
states using disposal options, 9*t.*
three strategic focuses, 27*f.*
treatment and disposal levels, 24*f.*
trends in, 10–11
for US municipal desalination plants, 10*f.*
concentrate pretreatment, 25
concentrate recycle, 61
Concentrate Reduction Facility (CRF), 183–184
concentrate reuse, 25
concentrate treatment, 14
conductor casing, 78
construction cost index, 121
construction costs
methods for estimating Class 4 and Class 5, 121–122
cost-capacity curves, 121–122
parametric cost curves, 122
vendor bids, 122, 123*t.*, 124*t.*
processes in desalination and concentrate management, 122–127
deep-well injection, 127, 127*t.*
evaporation ponds, 126–127
mechanical vapor compression (MVC) brine concentrators and crystallizers, 127, 128*t.*
microfiltration/ultrafiltration membranes, 123, 124*t.*

nanofiltration/reverse osmosis membranes, 123, 125, 125*t.*
pellet softener, 126, 126*t.*
solids-contact softener, 126, 126*t.*
construction methods, 145–146, 146*f.*
consumables, 128, 129*t.*
conventional pollutants, 74
cost-capacity curves, 121–122
costs, 31
criterion continuous concentration (CCC), 74
criterion maximum concentration (CMC), 74
crop yield, 89
Crystalactor® technology, 95
crystallization, 25
crystallizers, 110–111
influence of flow on operating costs, 111*t.*
MVC equipment cost, 128*t.*
process schematic, 110*f.*
current density, 57–58, 58*f.*
current leakage, 59f, 60

D

debt service cost, 127–128
decarbonators, 50
Decision H_2O, 19
deep-well injection, 13–14, 76–80, 146
advantages and disadvantages, 79–80, 79*t.*
Class I wells, 78*f.*
cost considerations, 79
current practice, 79
design considerations, 77–78
regulations, 76–77
degasifiers, 50
desalination, 1, 13, 24
brackish groundwater, 35–67
brackish water quality parameters, 15–16
cost, 119–130
source water quality affecting, 14–18
status, 2–7
total worldwide installed capacity by feedwater category, 3*f.*
total worldwide installed desalination capacity by user type, 4*f.*
US facilities, 4–7
desalting, 24
design-build procurement approach, 159–160
direct contact membrane distillation (DCMD), 100
direct coupled turbocharger, 43, 44*f.*
discharge options, 71
deep-well injection, 76–80
discharge to sewer, 75–76
discharge to surface water, 72–74
enhanced evaporation systems, 84–87
evaporation ponds, 80–83
irrigation, 87–91
disinfection byproducts, 163
drinking water
quality, 18–20, 19*t.*
source, 140
dust suppression, 201

E

electrical current efficiency, 60
electrical panels, 64
electrical staging, 56–57
ED systems, 56*f.*
membrane stack, 57*f.*
electrically driven membrane separation technologies, 26
electrodialysis, 26
membranes, 189–190
process schematic, 52*f.*
systems, 66–67
electrodialysis desalination, 51–65
chemical addition, 62
chemical cleaning, 62–63
chemical safety, 64
concentrate recycle, 61
current density and polarization, 57–58
current leakage, 59*f.*, 60
design considerations, 55–57
electrical current efficiency, 60
electrical operation, 64
electrode stream recycle, 62
mechanical systems, 65
membrane characteristics, 51–55
off-specification product recycle, 61
process description, 51
recovery rates, 61
safety considerations, 63
stack repair and disassembly, 63
electrodialysis metathesis (EDM), 26, 192
secondary, 98, 99*f.*
electrodialysis reversal (EDR), 26
computer models, 24*t.*
performance modeling, 23
phased reversal, 61
post-treatment, 67
pretreatment, 65–67, 65*t.*

process flow diagram, 97f.
reversal frequency, 62
salt removal rate, 58–59
secondary, 96–97
schematic, 96f.
SPARRO, 98–100
systems, 67
energy efficiency, 194–196
energy recovery, 42–44, 195
enhanced evaporation systems, 84–87
salt gradient solar pond, 85–87
spray systems, 84, 84t.
wind-aided intensified evaporation, 85, 85t.
enhanced recovery, 93–105
chemical softening, 94–96
chemical softening + secondary reverse osmosis/electrodialysis reversal, 97–98, 98f.
closed circuit reverse osmosis, 104–105, 104f.
forward osmosis, 100–101
High Efficiency Reverse Osmosis (HERO™), 102, 104f.
membrane distillation, 100
secondary EDR, 96–97
secondary EDR + SPARRO, 98–100
secondary electrodialysis metathesis, 98
Vibratory Shear Enhanced Process (VSEP®), 101–102
Enviro Water Mineral Co., 174
environmental issues, 141–152
equipment shutdown, 64
evaporation, 80
evaporation ponds, 14, 80–83, 133, 136, 177t.
advantages and disadvantages, 83, 83t.
cost considerations, 83
design considerations, 81–83
Ocotillo Brine Reduction Facility, 158
operation and maintenance, 83
regulations, 81
triple-line, for concentrate disposal, 178f.
evapotranspiration, 88

F

Faraday's law, 60
Fargo Wastewater Treatment Facility, Effluent Reuse Facility, 167–171
costs, 171
ERF water quality, 170t.
facility location map, 168f.
membrane system overview, 169–170
operations experience, 171
RO concrete management and permit, 170
RO unit parameters, 169t.
UF membrane characteristics, 169t.
wastewater reuse project schematic, 168f.
Federal regulations, 133–138
Florida Department of Environmental Protection (FDEP), 140
Florida membrane plants
discharge to surface water, 145t.
flux balancing, 42, 42t.
FMX technology, 192
forward osmosis, 100–101
"Frequently Asked Questions" (FAQs), 152

G

greensand, 49
Ground Water Rule, 163
grounding, 64

H

hard water, 19–20
High Efficiency Reverse Osmosis (HERO™), 31, 102, 104f., 194
OBRF process, 158
high-pressure pumping, 46–47
high-voltage AC power, 64
hot-mix asphalt (HMA) production, 202
hybridization, of desalination plant, 195
hydraulic loading rate, 88
hydraulic staging, 56, 56f.
hydrocarbons, 18
hydrochloric acid, 48, 62
hydrogen sulfide, 17, 66

I

individual permit, 134
informative tours/events, 151
injection casing, 78
inland desalination
overview, 1–11
public outreach and, 146–148
strategic focuses, 27f.
treatment and disposal levels, 24f.
innovations
design and operation, 190–192
membrane technology, 188–190
processes, 192–194
inorganic compounds, 16–18

integrated membrane systems, 49–50
integrated water balance model, 29
integrity testing, 44–45
interactive tools, 151
intermediate concentrate chemical stabilization (ICCS), 196–197
international regulations, 140–141
ion-exchange membranes, 53
ion selectivity, 51–52
ion toxicity, 144
iron, 16, 65
iron control, 48–49
irrigation, 87–91
  advantages and disadvantages, 90*t*.
Irwin Water Works, Army National Training Center, Fort Irwin, Calif., 164–167, 165*f*.
  costs, 166
  EDR concentrate management, 164–166
  EDR system design criteria, 166*t*.
  membrane system overview, 164
  process flow diagram, 166*f*.
  summary, 166–167

K

Kay Bailey Hutchison Desalination Plant, El Paso, Tex., 135, 171–174
  costs, 174
  desalination process schematic, 172*f*.
  membrane system overview, 172–173
  operations experience, 174
  plant location, 171*f*.
  public-private partnership, 174
  RO concentrate management and permit, 173–174
  RO trains, 172*f*.
  RO unit parameters, 173*t*.
  water quality, 173*t*.

L

labor cost, 129, 129*t*.
Laguna County Sanitation District Deep-Well Concentrate Disposal, 179–182
  concentrate management and permit, 180–181
  costs, 182
  facility location map, 179*f*.
  monitoring requirements for deep well, 182*t*.
  operations experience, 181
  RO system characteristics, 181*f*.
  treatment schematic, 180*f*.
  treatment system overview, 180
  WDR permit limits for discharge, 181*t*.
landfill, 7, 135
Langelier Saturation Index (LSI), 48
leaching fraction, 89
limiting current density, 59
location requirements, 144–145

M

m-phenylenediamine (MPD) and trimesoyl chloride (TMC) membranes, 188
manganese, 16, 65
manganese control, 48–49
maximum contaminant level (MCL), 19
mechanical systems, 47
media, 151
melt-blown cartridge filter, 48
membrane bioreactor system (MBR), system characteristics, 181*t*.
Membrane Brine Concentrator (MBC™), 101
  process schematic, 102*f*.
membrane distillation, 100, 141
  process schematic, 101*f*.
membrane flux, 38
membrane monitor (MeMo) detector, 190
membrane replacement, 128
membrane scaling monitoring, 190–191
membrane stack, 64
membrane technology, 188–190
microfiltration, 49–50
mixed matrix membranes (MMM), 188
MS2 virus, 45
municipal desalination facilities, 4–7
  cumulative installed capacity, 5*f*.
  number and types of plants, 5–6, 5*f*.
  number by state as of 2017, 6*f*.
  number of membrane processes as of 2017, 6*t*.
  number of plants installed by state in different time periods, 7*f*.
  plant locations, 6–7

N

nanofiltration, 16, 25–26
  versus reverse osmosis, 39
nanofiltration/reverse osmosis membranes, 123, 125, 141–144

construction cost
brackish water, 125*t*.
seawater, 125*t*.
innovations, 188–189
National Pollutant Discharge Elimination System (NPDES), 72
near-zero liquid discharge (NZLD), 8
nonconventional pollutants, 74
nonpotable reuse, 20–21

O

ocean outfalls, 141–144
off-specification product recycle, 61
online conductivity analyzers, 44
operation and maintenance
cost, 128–130
requirements, 146
organics, 18
osmosis, basic concept of, 37*f*.
osmotic pressure, 37–38

P

pan coefficient, 81
parametric cost curves, 122
particulate silica, 17
pellet-softening, 95
process, 95*f*.
pelletized lime softening, 196–197
water quality, 197*t*.
percolation, 88
permeability, 78
permeate, 36
pH, 142
adjustment, 144
phased reversal, 61
planning strategies, 28–32
combining technologies to form customized solutions, 30–31
new technologies, 31
tailored water quality, 31–32
polarization, 57–58
polarization point, 57
polyamide membrane, 36*f*.
porosity, 77
positive displacement devices, 43
post-treatment, 25
potable reuse, 21
California regulations for groundwater recharge into potable aquifers, 21*t*.
example of product water quality criteria, 22*t*.
potable water
quality, 18–20, 19*t*.
power cost, 129, 130*t*.
pressure-driven membrane, 25, 36–47
design considerations, 39–46
acid solution, 46
alkaline solution, 46
balancing flux, 42, 42*t*.
chemical cleaning, 45–46
energy recovery, 42–44
integrity monitoring, 44–45
membrane considerations, 39–41
propriety cleaning solutions, 46
recovery rates, 41
nanofiltration *versus* reverse osmosis, 39
osmotic pressure and reverse osmosis, 37–38
process description, 36
safety considerations, 46–47
structure of membranes, 36–37
treatment efficiency, 38–39
pretreatment, 24
proactive salinity management, 28–29
process flow management, 162–163
proprietary cleaning solutions, 46
public messaging, 150
public risk, 149
publicly owned treatment works (POTWs), 75–76, 134

R

reactive silica, 17
reclaimed water recycling, 32
recoverable salts, 200
potential application, 201*t*.
recovery rates, 41, 61
reference ambient concentration (RAC), 74
regulatory issues, 132–133
removal efficiency, 38, 39*t*.
renewable energy, 195–196
Resource Conservation and Recovery Act (RCRA), 133, 135–136
reverse osmosis, 16, 21, 25–26, 37–38, 116
advancement in membrane cleaning, 191–192
basic concept, 37*f*.
computer models, 24*t*.
membrane modules in membrane pressure vessel, 41*f*.
nanofiltration *versus*, 39
performance modeling, 23

post-treatment, 50–51
pretreatment, 47–50
reversible configuration, 191
RO+ZLD combined treatment cost, 113*t.*
treatment cost, 113*t.*
90% recovery, 114*t.*
two-stage process, 41*f.*
Reverse Osmosis System Analysis, 142

S

Safe Drinking Water Act (SDWA), 133, 135
salinity
in irrigation water, 88
seawater and membrane-filtered concentrates, 143*t.*
short-term goals, 20*t.*
salt gradient solar pond, 85–87
advantages and disadvantages, 87*t.*
schematic, 86*f.*
salt recovery, 196–202
pelletized lime softening, 196–197
significance and challenges of, 196
from sodium carbonate crystals, 199*f.*
techniques, 198–199
zero liquid discharge systems, 196
Salt Solidification and Sequestration (SAL-PROC™), 197–198
salt tolerance, 89, 90*f.*, 90*t.*
sand control, 48
saturation vapor pressure, 80
scale inhibitors, 30, 47
seawater desalination using RO (SWRO), 109
secondary maximum contaminant level (SMCL), 19
sewer discharge, 13, 75–76
advantages and disadvantages, 76*t.*
silica, 17–18
silt density index (SDI), 16
slurry precipitation and recycling RO (SPARRO)
secondary EDR and, 98–100, 99*f.*, 100*f.*, 194
social media, 151–152
sodium, 88
sodium adsorption ratio, 20–21, 89
sodium chloride, 63
sodium hypochlorite, 51
softening treatment facility, 158
soil remediation, 200
solids handling facility, 158
spacers, 54–55, 54*f.*
baffled flow pattern, 55*f.*
sparingly soluble salts, 17
Spiractor® technology, 95
spiral-wound configuration, 39, 40*f.*
state regulations, 138–140
California regulations, 139–140
Colorado regulations, 139
Florida regulations, 140
other local authorities, 140
Texas regulations, 139
statewide permit, 134
sulfuric acid, 48
supervisory control and data acquisition (SCADA) systems, 163, 164*f.*
surface casing, 78
surface water
advantages and disadvantages, 74, 74*t.*
design and cost considerations, 73–74
regulations governing discharge to, 72–73
suspended solids, 16

T

Tampa Bay Seawater Desalination Plant, 142
Tampa Bay Water, 145
Taunton River desalination plant, 50*f.*
Technologically enhanced naturally occurring radioactive material (TENORM), 133, 136–138
liquid residuals, 138*f.*
solid residuals, 137*f.*
Texas Commission on Environmental Quality (TCEQ), 139
thermal efficiency, 87
thermal evaporation, 26
thermo-ionic technology, 193
general schematic, 193*f.*
thin-film composite, 36–37
titanium oxide, 188–189
total dissolved solids (TDS), 15, 20
toxic pollutants, 74
toxicity, 142
ion, 144
transpiration, 88–89
turbidity, 65
turbocharger devices, 43
schematic representation of two-stage RO process, 43*f.*
two-pass configuration, 40

U

ultrafiltration, 49–50
  system overview, 162
    membrane design criteria, 162*t*.
ultrapure water, 2
Underground Injection Control (UIC) well system, 133, 135
underground source of drinking water (USDW), 135

V

Van't Hoff equation, 38
vendor bids, 122, 123*t*., 124*t*.
Vibratory Shear Enhanced Process (VSEP®), 101–102, 103*f*.
volume reduction, 25
vortex-based anti-fouling membrane system, 192

W

waste, form of, 134
  USEPA classification, 134*f*.
water quality, 13–32
  before and after pelletized treatment, 197*t*.
  analysis and monitoring, 21–23
  brackish water desalination technologies, 25–26
  concentrate management technologies, 26–28
  EDR, 167*t*.
  ERF, 170
  KBH, 173*t*.
  product water quality goals, 18–21
  RO and EDR performance modeling, 23
  source water quality affecting desalination, 14–18
  treatment levels, 23–25
water reuse, 13
water source sustainability, 148
  solutions, 149
Water Treatment Plant Number 2, 160
water treatment spectrum, 51, 52*f*.
websites, 151–152
whole effluent toxicity (WET), 73, 135, 144
wound cartridge filter, 48

Z

Zero Liquid Discharge Water Treatment Facility, 175–178
  brine concentrator system, 177*t*.
  concentrate management and permit, 177–178
  costs, 178
  evaporation ponds, 177*t*.
  falling film vapor compression evaporator, 176*f*.
  operations experience, 178
  RO system characteristics, 176*t*.
  treatment schematic, 175*f*.
  treatment system overview, 175–177
zero liquid discharge (ZLD), 8, 25, 105–117
  concentrate design flows, 163*t*.
  cost considerations, 112–116
  installations and institutional knowledge, 116–117
  process schematic, 161*f*.
  residuals management, 111–112
  RO+ZLD combined treatment cost, 113*t*.
    80% recovery, 114*t*.
    90% recovery, 114*t*.
    80% *versus* 90% recovery, 115*t*.
  supervisory control and data acquisition (SCADA) systems, 163, 164*f*.
  treatment, 106–111
  treatment cost, 112*t*.
  water quality parameters, 162*t*.

AWWA Manuals

M1, *Principles of Water Rates, Fees, and Charges*, #30001
M2, *Instrumentation and Control*, #30002
M3, *Safety Management for Water Utilities*, #30003
M4, *Water Fluoridation Principles and Practices*, #30004
M5, *Water Utility Management*, #30005
M6, *Water Meters—Selection, Installation, Testing, and Maintenance*, #30006
M7, *Problem Organisms in Water: Identification and Treatment*, #30007
M9, *Concrete Pressure Pipe*, #30009
M11, *Steel Pipe—A Guide for Design and Installation*, #30011
M12, *Simplified Procedures for Water Examination*, #30012
M14, *Backflow Prevention and Cross-Connection Control: Recommended Practices*, #30014
M17, *Fire Hydrants: Installation, Field Testing, and Maintenance*, #30017
M19, *Emergency Planning for Water and Wastewater Utilities*, #30019
M20, *Water Chlorination/Chloramination Practices and Principles*, #30020
M21, *Groundwater*, #30021
M22, *Sizing Water Service Lines and Meters*, #30022
M23, *PVC Pipe—Design and Installation*, #30023
M24, *Planning for the Distribution of Reclaimed Water*, #30024
M25, *Flexible-Membrane Covers and Linings for Potable-Water Reservoirs*, #30025
M27, *External Corrosion Control for Infrastructure Sustainability*, #30027
M28, *Rehabilitation of Water Mains*, #30028
M29, *Water Utility Capital Financing*, #30029
M30, *Precoat Filtration*, #30030
M31, *Distribution System Requirements for Fire Protection*, #30031
M32, *Computer Modeling of Water Distribution Systems*, #30032
M33, *Flowmeters in Water Supply*, #30033
M36, *Water Audits and Loss Control Programs*, #30036
M37, *Operational Control of Coagulation and Filtration Processes*, #30037
M38, *Electrodialysis and Electrodialysis Reversal*, #30038
M41, *Ductile-Iron Pipe and Fittings*, #30041
M42, *Steel Water-Storage Tanks*, #30042
M44, *Distribution Valves: Selection, Installation, Field Testing, and Maintenance*, #30044
M45, *Fiberglass Pipe Design*, #30045
M46, *Reverse Osmosis and Nanofiltration*, #30046
M47, *Capital Project Delivery*, #30047
M48, *Waterborne Pathogens*, #30048
M49, *Quarter-Turn Valves: Head Loss, Torque, and Cavitation Analysis*, #30049
M50, *Water Resources Planning*, #30050
M51, *Air Valves: Air-Release, Air/Vacuum and Combination*, #30051
M52, *Water Conservation Programs—A Planning Manual*, #30052
M53, *Microfiltration and Ultrafiltration Membranes for Drinking Water*, #30053
M54, *Developing Rates for Small Systems*, #30054
M55, *PE Pipe—Design and Installation*, #30055
M56, *Nitrification Prevention and Control in Drinking Water*, #30056
M57, *Algae: Source to Treatment*, #30057
M58, *Internal Corrosion Control in Water Distribution Systems*, #30058
M60, *Drought Preparedness and Response*, #30060
M61, *Desalination of Seawater*, #30061
M62, *Membrane Applications for Water Reuse*, #30062
M63, *Aquifer Storage and Recovery*, #30063

M65, *On-Site Generation of Hypochlorite,* #30065

M66, *Cylinder and Vane Actuators and Controls—Design and Installation,* #30066

M68, *Water Quality in Distribution Systems,* #30068

M69, *Inland Desalination and Concentrate Management,* #30069

M77, *Condition Assessment of Water Mains,* #30077

www.ingramcontent.com/pod-product-compliance
Lightning Source LLC
LaVergne TN
LVHW080652011026
849160LV00014B/1394